CAMEL TRACKS

# CAMEL TRACKS

## A HISTORY OF EXPLORATION, WARFARE AND POLICING IN THE MODERN IMPERIAL AGE

Andrew S. Goudie

Published in 2021 by:
The Society for Libyan Studies
c/o The British Academy
10–11 Carlton House Terrace
London SW1Y 5AH

www.societyforlibyanstudies.org



ISBN: 9781914268014



Cover and interior design by Chris Bell, cbdesign

Printed in the United Kingdom by Marston Book Services

# CONTENTS

# PREFACE

Over the years I have been fortunate to have been able to work in many of the world's drylands, including North and East Africa, Arabia, South Asia, China, Australia and the southwest USA. This has inevitably brought me into contact with one of Creation's most remarkable animals – the camel. In the past I have written about great desert explorers and about the role of the motor car in desert exploration. This book follows on from these previous works and instead relates the history of exploration *by camel* in the modern imperial era. It starts off by providing an overview of the camel in terms of their biology, their historical importance, their utility in the modern world, and the research that has been undertaken upon them.

I would like to add my thanks to the Publications Manager at the Society for Libyan Studies, Victoria Leitch, and the designer Chris Bell.

# CHAPTER ONE

# BACKGROUND

## BIOLOGY AND EVOLUTION

Camelids – the camel family – evolved in North America about 40 million years ago during the Eocene. One of these, called *Titanotylopus*, was the largest camel that has ever lived. It stood c.3.5 m high at the shoulder and ranged through Texas, Arizona, Kansas and Nebraska. It contrasted with *Protylopus*, which was about the size of a rabbit. Much later, camels spread to South America and Asia via the Panama Isthmus and the Bering Land Bridge respectively. Camels continued to inhabit North America until c.8000 years ago (i.e. in the Holocene, when they became extinct). In South America their descendants are llamas, alpacas, guanacos and vicuñas. Their modern Old World descendants are the one-humped dromedary (*Camelus dromedarius*) of North Africa and southwest Asia (**Figure 1.1**), and the two-humped Bactrian camel of central Asia. Bactrian camels are divided into wild (*Camelus ferus*) and domesticated (*Camelus bactrianus*) varieties. The term 'Bactrian' was first applied to two-humped camels by Aristotle, who noted in his *Historia Animalium* that 'The Bactrian camel differs from the Arabian in having two humps as against the latter's one'. One-humped and two-humped camels are capable of interbreeding. The name Bactrian relates to an area of Central Asia north of the Hindu Kush Mountains and south of the Amu Darya River in modern-day Afghanistan, as well as neighbouring parts of Tajikistan and Uzbekistan. Fossils of extinct camel species have also been found in the Old World – for example, in Syria (Martini 2019) and the Negev (Grigson 1983).

**Figure 1.1** A smiling dromedary in the Canary Islands. The wildlife photographer Cherry Kearton (1924, 97) described the appearance of the camel as 'stuck-up' and 'aristocratic-looking'. *Source: Author.*

The evidence of past camel hybridization – principally the crossing of Bactrian males and Arabian or dromedary females – from Central Asia in the east to Anatolia in the west (Inal 2020), is widespread. Hybrids appear to have greater strength and load-bearing abilities. However, although larger and stronger, hybrids look like dromedaries in that they have one hump, though this is normally not very symmetrical and can have a small indentation which divides the rear portion of the hump – often two to three times as large as the front – from the front part (Potts 2004). At the present time hybrid camels are favoured for camel wrestling bouts in western Turkey (Çaliskan 2009). Remarkably, studies of camel remains have shown that, in historical times, Bactrian, dromedary and hybrid camels were once widespread in Europe as far north as Britain, Belgium and Germany (Dioli 2018).

## DOMESTICATION

There are very small numbers of wild camels in the world today. The camel is a domesticated animal *par excellence*. Compared with dogs, which were already domesticated c.14,000 before the present (BP), pigs c.9,000 BP, cattle c.9,000 BP, sheep c.10,500 BP, goats c.10,000 BP, horses c.5,500 BP and donkeys c.5,000 BP, the domestication of the dromedary took place rather late in human history, most likely at the transition between the second and first millennia BC (Orlando 2016). Wild dromedaries must have occurred in considerable numbers on the Arabian Peninsula in Holocene times, as illustrated by the discovery of second millennium BC hunting sites along the eastern coast yielding large amounts of bones from this species, and through DNA and osteological studies of these bones (Uerpmann 2012; Almathen 2016). Zooarchaeological studies suggest that domestication first took place in the coastal southeast Arabian Peninsula.

Archaeological evidence indicates that the Bactrian camel was domesticated entirely independently of the dromedary in Mongolia and China around 5,000–6,000 years ago, possibly in Turkmenistan and Bactria (Sala 2017). By the 3rd millennium BC, they spread throughout much of Central Asia from China to Mesopotamia.

Georg Hartwig (1886), a popular natural historian of the Victorian era, referred (p.402) to the camel's 'thraldom', averred that 'its servitude is of older date, more complete and more irksome than that of any other domestic animal', that 'it bears all the marks of serfdom', and that they have suffered from 'an excess of misery and ill-treatment'. He continued (p.403), saying that 'The hardships of long servitude, which have thus gradually deformed the originally, perhaps not ungraceful camel, have no doubt also soured its temper, and rendered its character as unamiable as its appearance is repulsive'.

Even within the dromedaries, various different types evolved (Wardeh 2004) and they can be classified into four major classes: meat, dairy, dual-purpose and race camels.

## DISTRIBUTION

Extensive camel populations are very much a feature of a great belt of terrain stretching from West Africa across to Arabia and China, together with a large introduced feral population in Australia (**Figure 1.2**). The distribution of the domesticated Bactrian camels (red outline), wild Bactrian camels (black outline) and of dromedaries, (purple outline) is shown in Figure 1.2. The dromedary inhabits North Africa, the Horn of Africa and the Middle East, from the

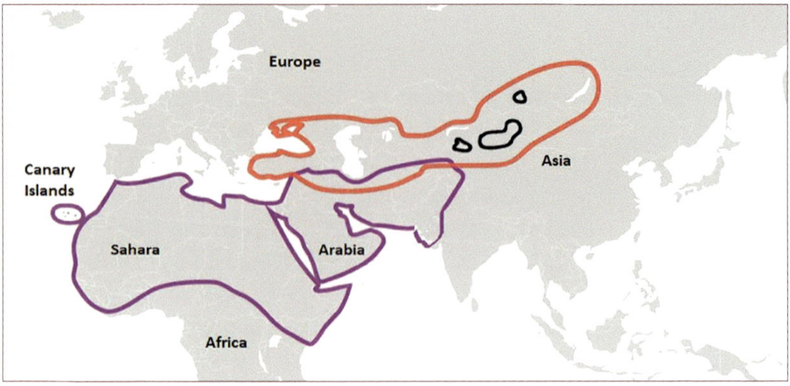


**Figure 1.2** The distribution of the dromedary (in purple), the Bactrian camel (in red), and the wild Bactrian camel (in black). In addition there are many feral dromedaries in Australia and a limited number in the USA. *Source: Author.*

Red Sea to the Indus; its wild ancestor is extinct. The Bactrian camel inhabits the cold deserts of Central Asia from the Black Sea to the Pacific Ocean, in the north of Iran and Afghanistan, in Middle Asia, Mongolia and the northern provinces of China. In Asia Minor and Central Asia east of the Caspian Sea, the ranges of the two types of camel overlap; it is here that hybrids are bred traditionally.

# HISTORY AND ARCHAEOLOGY

## HISTORICAL IMPORTANCE OF THE CAMEL

One of the reasons that one-humped camels were domesticated and then spread through most of North Africa, Arabia, Iran, Afghanistan, Baluchistan and north western India was that, as Buffon and Marsh indicated, they had a multitude of uses (**Table 1.1**, **Figure 1.3**). Indeed, they were so useful that over much of their geographical range the dromedary was utilized as a beast of burden in preference to wheeled carts. One consequence of this was that towns and cities evolved street plans that were to prove unsuitable for vehicular traffic. This is a theme that has been explored at length by Bulliet (1975). The extent to which these various uses were adopted varied from region to region, with, for example, the keeping of camels for milk in Somaliland, and the

**Table 1.1. Why are camels useful?**

| | | |
|---|---|---|
| Ambulances in time of war | Baggage carrying and beasts of burden | Blood for drinking |
| Blood money | Caravans of salt and other goods | Carrying guns |
| Cavalry in time of war | Dowry items | Excrement as fuel |
| Gelatin for catering | Geological surveying | Hair/wool for clothes, tents, yurts, etc. |
| Hunting | Jumping competitions | Leather |
| Manuring of fields | Meat | Milk, cheese, cosmetics |
| Milling of oil | Ploughing | Polo mounts |
| Portable libraries | Powder flasks made from scrotum and penis | Pulling carts |
| Pulling road rollers | Racing | Raising water from wells and irrigation channels |
| Sacrifice at religious festivals | Social status | Source of inspiration for art & poetry |
| Steeds | Tourist attractions (treks, safaris, rides, etc.) | Urine for cleaning, sal ammoniac, etc. |
| Water carrying | Wrestling/fighting competitions | |

**Figure 1.3** A heavily laden baggage camel and its Raika owner near Pushkar in Rajasthan, northwest India, in 1971. *Source: Author.*

use of camels for pulling carts by the Raika people in Rajasthan, India (Köhler-Rollefson 1992). Camels were also much employed as instruments of war and for policing the countryside. Some tribes in Africa used them for hunting ostrich (Parkyns 1850).

In addition, camels have an important role in traditional social relations in Somalia and Kenya, such as in payment of a dowry and in compensation of injured parties in clan feuds. In the Somali culture, camel ownership (in terms of herd size) is an indication of social status. Also, in the traditional economy of Somalia, camels are the main reserve stock and act as a store of wealth and security against drought, disease and other natural disasters (Guliye et al. 2007). In modern times, camels are a tourist attraction, as they are in the Canary Islands and at Petra (**Figure 1.4**), although they have been in the Canaries at least since 1405, where they have participated in virtually all agricultural work that required animal power. They also contributed to other different related activities, such as exploiting the steeper slopes to increase arable land, building retaining walls in terrace terrain or loading and transporting rocks where there was no alternative (Schulz et al. 2010, 562).

Camels helped in irrigation by raising water from wells. As Vita Sackville-West (1926, Chapter 2, Section 3) wrote:

> It is a relief to come upon a party of peasants at work in static attitudes, bent down over the earth, not walking on towards something else; with a camel near by, safely yoked, and turning the water-wheel from morning to night in the same trodden groove; this is a kind of triumph over the camel, which (with its outstretched neck) might be an animal designed to slouch onward, always at the same gait, always over the same desert, purpose subservient to pattern. A camel yoked is nature's design defeated, for the camel looks like a natural traveller, and not like a creature intended to walk round and round in the same circle. The wooden cogs squeal as they rub against one another, in the shadow of the tamarisk, and the little pitchers come up dripping out of the deep well, spilling half their water before it gets tipped out into the trench; a wasteful process, but one upon which the centuries have not been able to improve.

The value of camel meat and camel milk is now being reassessed. Camel meat contains less fat than lamb or beef, and its protein quality, assessed by the index of essential amino-acids in meat, is the highest among red meat. Its milk contains between 3 and 10 times more vitamin C than cows' milk (Abri and Faye 2019).

**Figure 1.4** Tourist camels at Petra, Jordan. *Source: Author.*

To enable camels to be ridden, and to carry baggage, they have to be provided with saddles. Many different types have evolved in different parts of the world. Further details are given in Baum (2015 and 2018). Women and children are sometimes provided with litters (Robinson 1931).

## CAMELS AND TRAVEL IN EXTREME ENVIRONMENTS

### *Adaptations*

Camels are wonderfully adapted to the environments in which they live. It is probably this that led the motor engineer Sir Alec Issigonis to quip in 1958 that the camel was a horse designed by a committee. They are large animals – c.2-m high at the shoulder – and so do not warm up or lose heat too quickly. Moreover, the adult dromedary's head is about 2.5 m above the ground, but with the neck stretched upward, it can reach the twigs and foliage 3.5 m from the ground that are inaccessible to most domestic animals and wild herbivores. They have large, padded feet, which are good for travelling through soft sandy terrain. Their long legs raise them above the highest temperatures close to the ground surface and also enable them to cover large distances. They have translucent eyelids that allow them to see tolerably well with their eyes shut, and nostrils that close to keep out sand. Their tough mouths and long incisors enable them to consume very thorny vegetation that is not palatable to many other animals.

However, the success of camels results from their ability to (1) tolerate elevated body temperatures, (2) minimize several avenues for water loss, and (3) reduce heat-gain from the environment.

A dehydrated camel can withstand body temperature fluctuations of up to 8°C – a fluctuation which would be lethal to most mammals. This tolerance spares camels from having to use unacceptably large quantities of water for keeping cool via evaporative cooling. An elevated body temperature decreases the temperature gradient between camels and their surroundings, thereby further reducing environmental heat gain. On cool evenings camels can transfer body heat back to the environment by radiation and convection. The camel's fur also has characteristics that reduce the chances of it becoming too hot. Camels can tolerate significant dehydration in comparison with other mammals. They can lose up to a quarter of body weight without ill effects. They can also rehydrate quickly when given the opportunity. A 600 kg camel can consume 200 litres of water in just three minutes. The camel's kidney is also designed for water conservation and can cope with very concentrated urine. They can also drink water of relatively high salinity and they also obtain some of their water requirements from the vegetation they consume. Their dung tends to be dry, so little water is lost through that means. Camels have a stomach structure that is different from other ruminants, but there is no firm evidence that this plays any role in water storage. Camels also have a hump or humps that contain fat which acts as a food store and permits camels to venture far from food sources for extended periods. The metabolism of the fat also yields water. Camels do not pant to lower their temperature and so do not lose water by that means. However, they do sweat, but only moderately. Camels can eat a wide range of foods, from dates and fish to thorny trees, date-kernels and salt marsh vegetation. Indeed, in Kutch, northern India, *kharai* camels swim through the mangrove swamps in search of food.

Although camels are generally hardy and healthy, they do suffer from various health problems, and some of these were noted by the explorers and military men who used them (Steel 1890; Leese 1927). Arnold Leese (1878–1956) worked as a vet in India and his book was a standard text for some decades. Unfortunately, his reputation became sullied in the 1930s and 1940s when he was a leading fascist and anti-semite. Among the diseases listed by Mukasa-Mugera (1981) are trypanosomiasis (*surra*), mange (spread by mites), ticks, mastitis, camel pox, anthrax, salmonella, tuberculosis and rabies. Some camel diseases can be transmitted to humans, including brucellosis and MERS (Middle East Respiratory Syndrome) (Zhu et al. 2019; Sazmand et al. 2019). MERS exposure to humans, which has become a major public health issue in the last decade, may be associated with consumption of unpasteurized camel milk, raw meat, viscera and urine. Camels in the United Arab Emirates are dying because of their propensity to consume plastic waste which then accumulates in their stomachs (Eriksen et al. 2021).

## CAMELS AND TRADE

The involvement of camels in long-distance transport of goods and people across arid regions was another highly important aspect of the usefulness of camels to humans. Many of the great desert explorers joined trade or pilgrimage caravans for their journeys. Some of these caravans were enormous, sometimes comprising thousands of camels. Angus Buchanan mentioned that the salt caravans from Bilma could contain as many as 7000 camels, whilst Dumreicher estimated that the date caravans from Siwa consisted of no less than 15,000 camels. The Blunts' caravan from Ha'il to Kerbela and Baghdad consisted of 4000 camels.

The dreadful trade in black slaves across the Sahara has been well documented by John Wright (2007) in his *The Trans-Saharan Slave Trade*. Likewise, the role of camels in trans-Saharan trade, particularly in Libya, has been discussed in Braun and Passon (2020), *Across the Sahara.*

Indeed, camels were an important facet of the world of the Garamantes in the Fezzan. They were probably present as a tribe in the Fezzan by 1000 BC. By around AD 150 the Garamantian kingdom in today's central Libya covered 180,000 square kilometres. It lasted from about 400 BC to AD 600. The Garamantes made use of elaborate underground irrigation systems known as foggaras in order to create a prosperous oasis in the heart of Sahara desert. They had several small planned towns and a capital, Garama (modern Germa/Jarma). They were involved with trade to both north and south, and for these camel caravans were crucial. Camel figurines have been found at Garama (Mattingly 2013).

Camels were used extensively in Roman times (Habinger et al. 2020). They had a multitude of uses, from racing to baggage carrying and ploughing. Their remains have been found from sites in Italy, the Iberian Peninsula, France, Belgium, Switzerland, Germany, England, Austria, Slovenia, Hungary, Ukraine,

**Figure 1.5** A pilgrimage caravan between Jedda and Mecca in c.1910. *Source: Kiernan 1937, 112.*

**Figure 1.6** Rendille men with camel sticks and white ear adornments in northern Kenya. The nomadic Rendille and Gabra people have societies and economies based on camels. *Source: Alice Goudie.*

Serbia and Bulgaria. Olwen Brogan (1954) reckoned that the camel was in Egypt and probably in Cyrenaica by the time of Augustus. The camel was sufficiently common in Tripolitania to be used for working the land in the Gefara in the second or early third century. Its role as a long-distance baggage animal may well have been established by the first century AD. Bovill (1956) suggested that somewhere between 46 BC and AD 363 the Roman army started using camels in North Africa.

Both the dromedary and Bactrian camel were imported to the northern Roman provinces and camels were present throughout the whole Roman period. While they may have been used for military purposes, they were also much used as pack animals (Pigière and Henrotay 2012). *Dromedarii* were camel-riding auxiliary forces recruited in the desert provinces of the Late Roman Empire in Syria. Both Bactrian and dromedary camels were used, and the first unit was formed by Emperor Trajan. They played a policing role.

The caravan trade was also important in parts of the Middle East, including Syria, where Palmyra was a major market (Seland 2015). Caravans were also used during the early and middle phases of the Byzantine Empire, as in Serbia (Marković et al. 2021).

## CAMELS AND PILGRIMAGE

Camel caravans were a vital feature of the annual pilgrimages to the Holy Places of Arabia (**Figure 1.5**). They assembled in places such as Cairo and Damascus and graphic descriptions are given in the works of Johann Burckhardt, Richard Burton, Joseph Pitt, Jack Keene and others. Jack Philby made the *hajj*, or pilgrimage, to Mecca in 1931, riding with the king and in the company of no less than 10,000 camels (Philby 1946). The caravan that Burton took to Medina was rather smaller, consisting of 7000 camels. Palgrave joined a caravan from Damascus that boasted 10,000 beasts.

## CAMELS AND NOMADIC SOCIETIES

There are many nomadic and semi-nomadic societies that are intimately linked to camel raising and usage, for example the Chaamba, Tubu and Tuareg of the Sahara, the Rgaybat of Western Sahara, the Beja of the Eastern Desert in Egypt, the Bisharin, Ababda, Kababish and Rezaigat of Sudan, the Turkana, Pokot (Suk), Gabra, Ariaal, and Rendille of Northern Kenya (**Figure 1.6**), the Borana of Ethiopia, the Afar people of Danakil, the Bedu of Arabia, the Raika and Rabari of north west India, and the nomads of Cholistan in Pakistan.

Camels enabled nomadic societies to be mobile (a *sine qua non* of nomadism) and allowed various economic activities to be carried out under highly adverse climatic conditions.

## CAMELS AND WAR

Camels have been used in war and in policing duties for many centuries (**Figure 1.7**). The camel was highly respected and used by the Persians, and its potential was exploited not only as a beast of burden but also as a pack animal in warfare. One of the earliest pieces of evidence of the use of camels in warfare comes from an Assyrian relief depicting a camel battle from Kalhu (Nimrud) Central Palace, and dating to 728 BC. Another early example of the military use of the camel is shown in a carving from the North Palace of Nineveh that depicts Assyrians pursuing Arabs on camelback and dated at around 660–650 BC.

At Persepolis (**Figure 1.8**) there are various Achaemenid carvings from c.500 BC which show camels and their handlers. The camel cavalry was first introduced into the Persian army by Cyrus the Great at the Battle of Thymbra. Herodotus stated that, at the Battle of Sardes in 540 BC, in which Cyrus defeated Croesus of Lydia, it was the camels that were responsible for the victory (Afshar 1978). In addition, the Ottoman Turks employed camels when they invaded the European Carpathian Basin in the sixteenth and seventeenth centuries (Bartosiewicz 2014). In 1590, at the direction of Ahmed El Mansur, ruler of Morocco, a blue-eyed Andalucian eunuch named Judar led a force with 8000 baggage camels across the Sahara to invade the Songhai Empire (Bovill 1927). Zamburak was a specialized form of self-propelled artillery on a camel with a mounted swivel gun (a small falconet), which was hinged on a metal fork-rest protruding from the saddle. It was employed by various regimes in Persia, and was also much used by the Mogul emperors in India between AD 1526 and 1858.

The use of camels for warfare in the Indian sub-continent has a long history. Camels were introduced in Sind (Sindh) in AD 717 by Muhamed Qasim, who had 3000 camels with him when he arrived to invade. The Afghan invader, Mahmud of Ghazni, passed through Jaisalmer in AD 997 and used thousands of camels to carry water to cross the desert. In the sixteenth century, Mughal Emperor Akbar, a contemporary of Queen Elizabeth I, and the maharanas of Rajasthan established camel corps for warfare (Köhler-Rollefson 2018). Akhbar was a great military innovator who consolidated Mogul rule across great swathes of India, including Baluchistan, Sind, Gujarat and Rajasthan.

**Figure 1.7** Assyrian camels, commanded by King Ashurbanipal, pursuing Arabs. The carving, made of gypsum, is currently housed in the British Museum. *Source: Wikimedia Commons.*

**Figure 1.8** Bactrian camel and coiffured handler (with hairband and earring) at Persepolis. *Source: Wikimedia Commons.*

**Figure 1.9** The Army of the Indus passing through the Bolan Pass on the border of Afghanistan. *Source: National Army Museum.*

He used camels to carry cannons. There were major camel routes that criss-crossed northwest India in Mugal times, linking places such as Agra and Surat (Anjum 2006).

In July 1798, during the Napoleonic Wars, the Army of the East, commanded by Napoleon Bonaparte, landed in Alexandria, Egypt with 30,000 soldiers. In January 1799, Napoleon established his own camel corps – the *Régiment des Dromadaires*. It was disbanded in 1801 after the French capitulation to the British. It was probably the first time that a European army used dromedaries in warfare.

A major acceleration occurred later in the nineteenth century when the colonial powers – Britain, France, Germany, Italy and Spain – developed camel corps. In the 1830s the British were concerned that the Russians might invade India via the Khyber and Bolan passes that lay on the borderland between Afghanistan and India. British military men and explorers, such as Alexander Burnes, were active gathering intelligence in Afghanistan, Persia and Bokhara, often using camels. As part of an attempt to deal with the perceived Russian threat, the First Anglo-Afghan War was fought between the British Empire and the Emirate of Afghanistan from 1839 to 1842. During this war, the 'Army of the Indus', which included 21,000 British and Indian troops, set out from Punjab in December 1838 (see Morrison 2014) (**Figure 1.9**).

This army boasted more than 30,000 camels. By the time the column reached Kandahar it was estimated that more than two-thirds of these had been lost. One of the reasons for this was that the camels were overloaded, as some of the officers had extravagant tastes and refused to travel light. As William Dalrymple (2013, 152–3) pointed out, one brigadier needed 50 camels to carry his kit; General Cotton needed 250; 300 hundred camels were earmarked to carry the military wine cellar; and one regiment employed two camels just to carry the best Manila cigars.

In 1842, the British, under Sir Charles Napier, conquered the province of Sind, located in the south of modern-day Pakistan. On Napier's instructions, the Scinde Camel Corps (SCC) was raised in December 1843 at Karachi. It was short-lived, as in 1853, it was designated as the 6th Regiment of Punjab Infantry and became part of the Punjab Irregular Force. Nearby Bikaner, a princely state in the Rajasthan area of northwest India, is located in the Thar Desert. The Bikaner Camel Corps (BCC) was founded there by the Maharaja Ganga Singh (1880–1943) in 1889 as the 'Ganga Risala' after the government of India accepted his offer to raise a force of 500 soldiers. Risala is an Urdu word meaning 'mounted cavalry'. Ganga Singh led the Ganga Risala when it fought in the Boxer Rebellion in China in 1900, in Somaliland in 1902–1904, and in Egypt in the First World War. In 1915, at the Suez Canal, the BCC routed the opposing Turkish forces in a camel cavalry charge.

General Robert Napier used 7000 camels during the punitive Abyssinian Expedition of 1868. In 1888, after signing a succession of treaties with the then ruling Somali sultans, the British established a protectorate in northern present-day Somalia, referred to as British Somaliland. Various British hunting and archaeological parties used camels to explore the country at this time (Parkinson 1898; Seton-Karr 1898; Pease 1898). Sir Henry Seton-Karr found so many palaeoliths that he over burdened his camels and they struggled to reach the coast. The Somaliland Camel Corps (SCC) was a unit of the British Army based in the protectorate. It lasted from early 1914 until 1944. The British used camels in the Sudan, somewhat unsuccessfully, in an attempt to relieve General Gordon (**Figure 1.10**) at Khartoum during the Nile Campaign of 1884–5.

In the Sahara, François-Henry Laperrine (1860–1920), sometimes touted as the French equivalent of Lawrence of Arabia, was a great soldier who, over four decades, helped to establish French rule in North Africa. Laperrine drew his African recruits mainly from the Chaamba (nomads of the Ouargla district of southern Algeria), organizing them into three 300-man

**Figure 1.10** General Gordon en route to Khartoum. *Source: Anon.* Pictorial Records of the English in Egypt*, undated.*

camel companies led by French officers. The three units of *Méharistes* (camel cavalry) were responsible for three different areas of the Sahara – Salah, Timimoune and Touat. They proved to be cheap and highly professional. The Chaamba, like the Tuareg, were masters of the desert. They were tough, knew the desert intimately, had great tracking and navigational skills, and were happy to attack the Tuareg, with whom there had been long-standing animosity.

The Italian colonial regime in Libya recruited a camel corps for use in Cyrenaica and Tripolitania. This *Zaptie Méharista* served primarily as desert gendarmerie. Like their French counterparts, they were recruited from the indigenous desert tribes and wore modified versions of tribal dress. On the Atlantic coast between Morocco and Mauritania, the Spanish had a colony called Sahara. The *Tropas Nómadas* (Nomad Troops) were an auxiliary regiment of the Spanish army from the 1930s until the end of the Spanish presence in the territory in 1975. Composed of Sahrawi tribesmen, they were equipped with small arms and led by Spanhish officers. They guarded outposts, served as desert police and conducted patrols on camelback. The corps was modelled on the French *Méharistes*.

In southern Africa, lies Namibia (formerly Southwest Africa). The Germans invaded Southwest Africa in the 1880s, fought horrendous battles with the local people and remained until they were defeated in 1915 by South African troops during the course of the First World War. The German military forces, called the *Schutztruppe* (Protection Force), made extensive use of camels in their bid to take control of this arid country and to enable much of the land to be taken over by German settlers. Camels were also used in the drier parts of the Cape Colony (from 1897), in Southern Rhodesia (from 1903) and in Northern Rhodesia for the transport of mail and for the patrols of the British South Africa Police. Camels were also used for policing duties in Bechuanaland (Botswana) (**Figure 1.11**).

T.E. Lawrence (1888–1935), Lawrence of Arabia, is famous for his campaigns in the First World War against the Ottoman Turks in Sinai and in Palestine. With the Arab Army he used camels to good effect, although he was not averse to using armoured cars and aircraft when necessary. He operated in the hills and stupendous cliffs of Wadi Rum (**Figure 1.12**). L. Thomas (1924, 146), his great admirer and publicist, averred that Lawrence was the only European he had ever met, who possessed 'camel instincts' – 'a quality that implies intimate acquaintance with the beast's habits, powers and innumerable idiosyncracies'. At the end of 1915, Lieutenant-Colonel Leslie Smith VC, aware of the problems of warfare in waterless deserts, approached General

**Figure 1.11** Extremely smart Botswana Police with their camels at Tsabong, southern Kalahari, in 1967. *Source: Author.*

**Figure 1.12** The cliffs and stupendous hills of Wadi Rum. *Source: Author.*

Sir John Maxwell, who commanded the Egyptian Expeditionary Force (EEF), for permission to raise a force of camel cavalry. Such a force would have value in combatting the Senussi in the Libyan Desert and the Turks who threatened the Suez Canal. In due course, the Imperial Camel Corps (ICC) was raised in December 1916. It grew to a brigade of four battalions, one battalion each from Great Britain and New Zealand and two battalions from Australia.

Although the military use of camels is now negligible, they are still used for police duties in various countries. For instance, in northern Kenya a new police unit, based on Isiolo, has recently been unveiled to patrol remote sections of the Kenya-Somali border in efforts to combat banditry.

## CAMELS IN THE MODERN WORLD

### AUSTRALIAN CAMELS

One fascinating aspect of the story of camels and their distribution, is that of their introduction to Australia and the USA in the mid-nineteenth century. The first camel, a dromedary bull named Harry, arrived in Australia from the Canary Islands in 1840. This camel, said to be bad-tempered and prone to biting people, was used by John Horrocks (1818–1846), a settler and explorer, on an expedition beyond the Flinders Ranges in 1846. When Harry lurched, one day while out walking, Horrocks' gun discharged and he was mortally wounded, dying 23 days later.

The most important development in the use of the camel in Australia came with the ill-fated Burke and Wills expedition. Another turning point occurred in 1866 when a land magnate, Thomas Elder, imported camels from Karachi, established a stud at Beltana in Southern Australia and loaned them to expeditions such as that led by Peter Warburton. Elder also imported cameleers to manage the camels. These men were mainly referred to as 'Afghans', though in reality many of them came from Rajasthan, Baluchistan and other parts of British India.

Major expedition using camels were led by Ernest Giles, William Gosse, William Tietkens, David Lindsay, William Horn, David Carnegie, Larry Wells, Alfred Carnegie and Cecil Madigan. The commercial use of camels then expanded.

**Figure 1.13** An 'Afghan' camel driver with a camel train loaded with chaff, c.1911.
*Source: Photograph courtesy of the State Library of South Australia. SLSA: B 14808.*

## USA CAMELS

For a short period in its history, camels became an important issue in the United States (Gray et al. 1930; Fabens 1865). To achieve the use of camels involved a long, hard fight for Jefferson Davis, which began in 1851 when, as a member of the United States Senate Committee on Military Affairs, he had introduced the first resolution providing an appropriation to purchase camels for experimentation. After almost four years of pleading, cajoling and debating, Congress was finally stirred in 1854 to appropriate $30,000 for the project. Davis' opinions about the feasibility of a mounted desert corps were in part moulded by G.P. Marsh, who had travelled extensively in Egypt, Syria and Arabia, and had long studied the possibility of such a venture. Bactrian camels were introduced (**Figure 1.14**), but on a smaller scale than the dromedary.

Edward Fitzgerald 'Ned' Beale (1822–1893) demonstrated the value of camels when he set out westward from Fort Defiance in New Mexico to Fort Yuma on the Colorado River border of Arizona and California with 25 of the beasts. Beale's Wagon Road, 1600 km long, was to be followed in part by Route 66. Beale was pleased to note that the camels swam the Colorado with relative ease. He employed a camel driver, Hadji Ali (known locally as Hi Jolly). In April 1858, Beale was ordered to survey a second route along the thirty-fifth parallel from Fort Smith, Arkansas, to the Colorado River for use as a wagon road and stage line. Camels were also used in Death Valley in 1861 to delimit the border between Nevada and California (Woodward 1961).

However, the so-called camel experiment did not last long. The arrival of the railroad and the disruption caused by the American Civil War led to its demise.

Baum reported (2011, 77):

> . . . a little over two thousand Arabian camels and three to five-hundred Bactrian camels reside in the US. Many are zoo or circus stock, but most are privately held by perhaps as many as twenty or thirty individuals who breed, sell or work camels. Herd numbers range between eight and eighty camels and significant numbers dwell piecemeal around the country on private game ranches.

## NUMBERS

Compared to other domesticated animals, such as cattle, sheep, goats and pigs, camels are relatively few in number. It is difficult to get a clear picture of the global camel population (Faye 2020), but it is probably around 30 million head. Approximately 90–95% are dromedaries. Bactrian camels are much rarer. There are only around one million of them, nearly all of which live in China, Mongolia, Kazakhstan and Russia. For comparison, the global cattle and sheep populations both exceed one billion. The camel population represents less than 1% of the total herbivorous domestic population in the world (Faye 2014).

**Figure 1.14** American sailors loading Bactrian camels onto the USS *Supply* in 1856 for importation into the USA. *Source: plate in Jefferson Davis' Report to the Secretary of War.*

FAO data indicates that camels occur in large numbers in certain specific countries. Somalia and Chad both have in excess of 7 million head of camels. Other countries with more than a million camels include Ethiopia, Kenya, Mali, Mauritania, Niger, Pakistan and Sudan. Although in the popular mind camels are often associated with Arabia, in reality over 80% of the world's camel population lives in Africa.

The numbers of camels has been increasing over recent decades in some countries. For example, in 1968 Somalia had 4.6 million camels compared with 7.2 million in 2018. Conversely, in many countries, the number of camels has declined. India, for example, had 1.08 million camels in 1968, but only 0.3 million in 2018. Libya had 0.23 million camels in 1968, but only 0.07 million in 2018. Today, there are approximately 4000 camels in the Negev, i.e. about two camels per 100 Bedouin, whereas in 1955–1961, there were between 56 and 61 camels per 100 Bedouin (Degen et al. 2019). In Turkey, before the Second World War, there were over 100,000 camels, but by the start of the twenty-first century the figure was c.1000–2000 (Yilmaz et al. 2011). Camels as means of transport, as beasts of burden and for ploughing, have often been replaced by motor vehicles, but camel populations worldwide have been increasing, mainly due to demand for camel meat and milk (Faye 2015). Faye (2020) calculated that between 1961 and 2018, the world camel population multiplied by 2.75, a higher value than equines (1.06), sheep (1.21), cattle (1.58), small camelids (1.72) and buffalo (2.33). Only the growth of the goat population appears higher (3.00). Camel meat is reputed to be healthy, being low in cholesterol, and there is much interest in the health-giving properties of camel milk, and in the possible medicinal potential of camel urine. Additionally, because they can survive and produce milk under very arid conditions, camels are being added to the herds of many pastoralists in increasingly drought prone areas, notably in Eastern Africa where they are replacing cattle. This is the case with the Borana of northern Kenya (Kagunyu and Wanjohi, 2014).

In Australia, there are large numbers of feral camels. As we have already seen, the dromedary was introduced in 1840, and was used by early explorers such as Warburton. Between 1880 and 1907 some 20,000 were imported. However, with the advent of motor transport they became largely redundant, and between 1920 and 1941 between 5000 and 10,000 were released into the wild. There are now around one million feral camels in Australia, spread across 3.3 million km$^2$. Their effects on aboriginal communities can be devastating. The presence of so many feral camels gives a lie to the oft stated view that camels are incapable of mating without the assistance of their handlers.

There are very few truly wild camels left in the world, though there are some wild Bactrians (*Camelus ferus*) still present in Central Asia. Critically endangered, they live in parts of northern China and southern Mongolia, but only about 1000 camels still survive. Most live on the Lop Nur Wild Camel National Nature Reserve in China, and a smaller population lives in the Great Gobi Protected Area in Mongolia. As John Hare recognised, their principal enemy is man, who hunts them, prospects for oil, undertakes nuclear tests and poisons their grazing land with potassium cyanide in the search for oil. Xue et al. (2021) provide a picture of their current distribution.

# HISTORY OF RESEARCH

## INTRODUCTION

There are many useful books on camels, recent examples include *The Camel* (1981) by Gauthier-Piltrers and Daggs, and *Camel* (2010) by Robert Irwin. The former covers camel evolution, ecology and behaviour, together with domestication, nomadism and uses. Irwin's less conventional but engaging treatment deals *inter alia* with camels in literature and art and their relationship to Islam. Another useful work is *Camel* by Wilson (1984). This deals with the one-humped camel, describing camel origins, distribution and functions, along with management, anatomy, physiology and health. A particularly useful source of bibliographic information is provided by Mukasa-Mugerwa (1981).

There are three earlier works that still have considerable value: those by Leonard, Buffon and Marsh (see below). Major Arthur Leonard's *The Camel – Its Uses and Management* appeared in 1894. Leonard (1856–1909?) was a soldier who was born in Ireland but went on to become a Rhodesian pioneer. In 1873 he joined the 2nd Battalion, East Lancashire Regiment, serving as a transport officer in India and in conflicts in Afghanistan, Egypt and the Sudan, where he was seriously injured during the Mahdhist revolt of 1885. His book is essentially a practical guide to such issues as breeding, feeding, watering, loading, marching, ailments, equipment, management and purchasing, and their multiple advantages over ox transport. He also discusses the temperament of camels, using adjectives such as passive, long-suffering, serious, gloomy, lugubrious, submissive, tractable, stupid, plucky, vindictive and passionate when aroused. On the other hand he stresses their utility, especially when well-handled.

## COUNT BUFFON

A major figure in the early study of camels was Georges-Louis Leclerc, Comte de Buffon (1707–1788). Hailing from Burgundy, he became one of the greatest natural historians of the age. In 1739 he was appointed head of the Parisian Jardin du Roi, a position which he held to the end of his life. Buffon's greatest contribution to knowledge was his monumental, multi-volume *Histoire Naturelle*. This was translated into English in ten volumes. Volume 8, published in 1797, starts with a lengthy discussion of camels and was one of the first scientific treatments of these animals. It contained two representations of these exotic beasts.

Buffon described the two types of camel (**Figure 1.15**) and discussed their distribution (p.2). I have not sought to modernise his English:

> Aristotle has judiciously marked the two principal races; the first, which has two hunches, under the name of the *Bactrian Camel*; and the second under that of the *Arabian Camel*: the first are called *Turkish* and the others *Arabian Camels*. This distinction still subsists, but it appears, since the discovery of those parts of Africa and Asia which were unknown to the ancients, that the dromedary is, without comparison, more numerous and more universal than the camel: the last being seldom found in any other place than Turkestan, and some other parts of the Levant; while the dromedary, more common than any other beast of burthen in Arabia, is found in all the northern parts of Africa, from the Mediterranean to the river Niger, and is also met with in Egypt, in Persia, in Southern Tartary, and in all the northern parts of India. The dromedary, therefore, occupies an immense tract of land, while the camel is confined to narrow limits. The first inhabits hot and parched regions; the second, a more moist soil and temperate climate...

Buffon noted their utility to the Arabs (p.5):

> Blest with their camels, they not only want for nothing, but they even fear nothing. In a single day they can traverse a tract of fifty leagues into the desert, and thus escape from their enemies. All the armies in the world would perish in pursuit of a troop of Arabs; and hence they are no further submissive than they please.

He also discussed their load-carrying capabilities and their speed (p.9):

Large camels generally carry 1000, or even 1200lbs weight, and the smaller 6 or 700. In these commercial journeys, they do not travel quick, and as the route is often seven or eight hundred leagues, they regulate their motions and journeys; they only walk, and go every day ten or twelve leagues; they are unloaded every evening, and are suffered to feed at liberty. In a country where there is plenty of pasture, they eat enough in one hour to ruminate the whole night, and to serve them twenty-four; but they seldom meet with such pastures, and this delicate

**Figure 1.15** Buffon's illustrations of the Bactrian (top) and dromedary (bottom) camels. *Source: Buffon 1797, volume 8, Frontispiece.*

> food is not necessary for them. They even seem to prefer worm-wood, thistles, nettles, furze, and other thorny vegetables to the softest herbs; and as long as they can find plants to brouze on, they easily dispense with drink.

He was also interested in how the camel had been transformed by human actions (p.10):

> If we compare the deformities, or rather the non-conformities of the camel with other quadrupeds, we cannot doubt but his nature has been considerably altered by constraint, slavery, and continual labour. The camel is the most completely, the most laboriously, and the most anciently enslaved of all domestic animals; the most anciently, because he inhabits climates where man was the most early civilised; the most completely, because in the other species of domestic animals, such as the horse, the dog, the ox, the sheep, the hog, &c. we find some individuals in their natural state, which have not yet been subjected by man; but the whole species of the camel is enslaved, and not any of them are to be found in their primitive state of independence and liberty; and lastly, he is the most laborious slave, because he has never been trained, either for shew, as are many horses, or for amusement, as are almost all dogs, or for the use of the table, as are the ox, the hog, the sheep, &c.

In a way that prudish Victorian Englishmen might have found difficult, he discussed camel copulation and genitalia (p.17–18):

> The camel does not copulate like other quadrupeds, for the female sinks upon her knees and receives the male in the same situation as she rests, sleeps, or is loaded. This posture, to which they are easily accustomed, becomes natural to them, since they assume it at the time of their copulation. The female goes about twelve months with young, and, like all large quadrupeds, produces but one at a birth: they have great plenty of milk, which is thick and nourishing, even for the human species, when mixed with a great quantity of water. The females seldom do any labour when with young, but are suffered to bring forth at liberty.
>
> The genital member of the male, like that of the bull, is very long and slender; it tends forward during copulation, like that of every other animal; but in its usual state, it is bent backwards, and voids the urine between the legs, so that the male and female pass their urine in the same manner.

Finally, Buffon re-iterated the value of the camel (p.17–18)|:

> Gold and silk are not the true riches of the east, the camel is the treasure of Asia.
>
> The camel is not only of greater value than the elephant but perhaps not of less than the horse, the ass, and the ox, when all their advantages are united. He carries as much as two mules, and not only eats less, but feeds on herbs as coarse as the ass. The female furnishes milk longer than the cow; the flesh of young camels is as good and wholesome as veal; their hair is finer, and more sought after than the best wool. Even their excrements are useful, for sal ammoniac is made of their urine, and their dung, when dried and powdered, serves them for litter, as well as the horses, with whom they often travel in countries where neither straw nor hay is known. To conclude, they also make excellent fewel of this dung, which burns freely, gives a flame as clear, and almost as lively, as that of dry wood, and which is of great use in the deserts, where not a tree is to be seen, and where, from the deficiency of combustible matters, fire is almost as scarce as water.

## GEORGE PERKINS MARSH

Another nineteenth-century work to which one should turn, is that of George Perkins Marsh (1802–1882). He was a pioneering conservationist and ecologist from Vermont. A true polymath, he served as a congressman and as American envoy to Turkey and Italy. In 1864, he wrote his greatest work, *Man and Nature*, which was the first book to explore in depth how humans were transforming the Earth. This interest in the human impact inevitably led to his having an interest in domestication, and that in turn caused him to take an interest in the camel. In 1855 he delivered a lecture (which was also published) on the camel to the Smithsonian Institution in Washington, D.C. This was followed in 1856 by a full length book *The Camel: His Organization, Habits and Uses Considered with Reference to His Introduction into the United States*.

In this book, Marsh stresses the role that the camel has played in opening up some of the driest parts of the world (p.21–2). He also summarised the peculiarities of the camel that rendered it so useful to humans (p.37):

> The general anatomy of the camel is the same as that of other ruminants; but the hump, the horizontal posture of the head, the direction of the eye, the power of closing the nostril, the callosities upon the breast

> and legs, the spreading and cushioned feet, and above all the curious structure of the stomach, to which he owes his most valuable property, the power of long abstinence from water, distinguish him from all other quadrupeds.

However, he was particularly interested in the value that the camel might have in the United States and its potential role in military endeavours. He wrote appreciatively on the spectacle that advancing camels could present and the influence that this might have on the native peoples of the Plains (p.188):

> There are few more imposing spectacles than a body of armed men, advancing under the quick pace of the trained dromedary; and this sight, with the ability of the animal to climb ascents impracticable to horses, and thus to transport mountain howitzers, light artillery, stores, and other military material into the heart of the mountains, would strike with a salutary terror the Comanches, Lipans, and other savage tribes upon our borders.

## SCOPE OF THE BOOK AND ITS MAIN THEMES

Camels are beguiling creatures that have played a major role in the history of exploration, in the economies and societies of the environments in which they are found, and in warfare and policing. This book explores these three themes, based largely on a study of the accounts of European explorers and soldiers operating in North Africa and the Middle East. The book does not cover the explorations that took place further east in Asia using Bactrian camels, nor those of the interior of Australia. The remaining chapters of this book are essentially concerned with the use of camels for exploration, largely by Europeans, in the Modern Imperial Era. It gives details of the lives of the explorers and also gives their diverse and often divergent views on the virtues and vices of camels.

It is important to remember, however, that much of the European exploration of 'empty' spaces in Africa, Arabia and Asia capitalized on already established camel networks. Without the camel and the people who were already exploiting the remarkable qualities of camels in difficult ecological zones, much of the history described in the coming pages would not have been possible. Particularly in the eighteenth century and the first part of the nineteenth century, many of the explorers, such as Browne, Niebuhr, Tavernier, Hornemann,

Caillié, Burton, Doughty, Palgrave and the Blunts travelled with long-distance trading or pilgrimage caravans.

I have, by use of extensive quotations, tried to give a picture of how camels have contributed to exploration and military and policing activities, and to record some of the great and heroic journeys that have been undertaken with them. I have not attempted to omit quotations that contain offensive, racist or sexist remarks about local people. Regretfully, such remarks, which I deprecate, were a feature of some of the explorers discussed in this book. Some of the explorers, however, had a great respect and liking for the people they encountered. Their views too are given. Readers may also note a gender imbalance. The reality is that most of the explorers were men, but I have deliberately drawn attention to some remarkable women, including Mabel Bent, Freya Stark, Alexine Tinne, Anne Blunt, Florence Baker and Gertrude Bell.

Camels played a major role in the history of exploration of the Earth's drylands. This is exemplified by a consideration of the number of medals that those who employed them have received from the Royal Geographical Society (RGS) (**Table 1.2**).

There were of course many earlier explorers who used camels and accompanied caravans across North Africa and Arabia. Pride of place should perhaps go to the Moroccan traveller Ibn Battuta (1304–1368/9). Over a period of c.30 years he travelled some 100,000 km, taking in China, South-east Asia, India, the Swahili Coast of East Africa, Iberia, Arabia and the Sahara. Also of note was the Andalusian berber Leo Africanus (1494–1554). Amongst his many travels, he went to Arabia and crossed the Sahara with a caravan of camels.

**Table 1.2. Camel-Riding Gold Medallists of the RGS (1850–1960)**

| | | |
|---|---|---|
| **1960** Theodore Monod | **1924** Ahmed Hassanein | **1898** Sven Hedin |
| **1957** Ardito Desio | **1920** Harry St John Philby | **1882** Gustav Nachtigal |
| **1948** Wilfred Thesiger | **1918** Gertrude Bell | **1874** Peter Warburton |
| **1942** Freya Stark | **1918** Jean Tilho | **1868** Gerhard Rohlfs |
| **1942** Owen Lattimore | **1912** Charles Doughty | **1865** Samuel Baker |
| **1936** George Murray | **1912** Douglas Carruthers | **1862** Robert O'Hara Burke |
| **1931** Bertram Thomas | **1911** Pyotr Kozloff (Kozlov) | **1859** Richard Burton |
| **1929** Francis Rennell Rodd | **1909** Aurel Stein | **1851** Georg Wallin |

Some explorers attributed to camels their desire to travel in desert climes. Freya Stark (1893–1993), undoubtedly one of the greatest lady travellers in Arabia during the twentieth century, was in Damascus when her first sight of beautiful camels converted her to a long desert career (*Letters from Syria*, 1942, 109):

> And then the wonder happened! Camels appeared on our left hand: first a few here and there, then more and more, till the whole herd came browsing along, five hundred or more. I got out and went among them to photograph. The two Beduin leaders, dressed gorgeously, perched high up and swinging slowly with the movement of their beasts, shouted out to me, but the Beduin Arabic is beyond me. I can't tell you what a wonderful sight it was: as if one were suddenly in the very morning of the world among the people of Abraham or Jacob. The great gentle creatures came browsing and moving and pausing, rolling gently over the landscape like a brown wave just a little browner than the desert that carried it. Their huge legs rose up all round me like columns; the foals were frisking about: the herdsmen rode here and there. I stood in a kind of ecstasy among them. It seemed as if they were not so much moving as flowing along, with something indescribably fresh and peaceful and free about it all, as if the struggle of all these thousands of years had never been, since first they started wandering. I never imagined that my first sight of the desert would come with such a shock of beauty and enslave me right away. But I left it feeling that somehow, some time, I must see more of the great spaces.

However, she seems to have done the bulk of her travels on other steeds, including mules.

Some of the explorers who used camels were serious scientists. This is the case with respect to people like Chudeau in the Sahara (see this volume), Frederic Cailliaud in the Libyan Desert (Cailliaud 1822), and Beadnell in Sinai (see Beadnell 1909). Of particular note in Libya was the Italian geologist Ardito Desio (1987–2001) (see Desio 1950; Orombelli 2004). Like Chudeau he was greatly responsible for the finding of oil in northern Africa. In 1926, Desio organized and led an expedition to Jaghbub oasis. From 1930 to 1933 he led expeditions through the Libyan hinterland, including the crossing of the Sahara with a large caravan of camels from the Mediterranean seaboard as far as the frontier of Sudan and back through Fezzan. In 1935 and 1936 he further explored the Fezzan and the Tibesti Massif in the Central Sahara. From 1936 to 1940, he organized and directed the Libyan Geological Survey. In 1938 Desio discovered oil in the sub-surface of Libya. In Persia, J.V. Harrison used camels to explore for oil on the Makran Coast (Harrison 1941).

Some explorers who used camels were archaeologists: Giovanni Battista Belzoni in Egypt (Belzoni 1835), Captain Robert Murdoch Smith and Commander Edwin Porcher (1864) in Cyrenaica, and Gertrude Bell in Arabia. Some took a great interest in anthropology (e.g., Luigi Robecchi Bricchetti in Somalia). Some others were missionaries (e.g., Samuel Zwemer in Arabia and Lewis Krapf in Eastern Africa), or sought to explore sites of religious significance in the Holy Land (e.g., David Roberts and Henry Field). Some sought to counter slavery (e.g., the Bakers in East Africa). Still others were big game hunters, as in East Africa (e.g., F.B. Parkinson and James Harrison) and Central Asia (e.g., Nicolay Przhevalsky), or wildlife photographers (e.g., Cherry Kearton in Algeria). Boundary delimitation and surveying was another motive (e.g., George Tate in Seistan). Quite a number of them were involved in espionage.

European travellers did not always find riding on camels to be easy. Ranulph Fiennes, the accomplished military officer and explorer, described the discomfort of his privates in Oman (2015, 130):

> I had ridden horses on and off for years but they are well designed to be ridden bareback by humans with well-cushioned backsides. Camels are not, and their spines behind the hump jar their rider's coccyx painfully until they get the knack of how to be comfortable, if contorted, whether by side-saddle riding or, like the bedu, literally kneeling on the upturned soles of your feet.
>
> For my first few rides I felt as though my privates were being tortured, so I switched to the lesser evil of bruised buttocks.

We will now turn in chapter 2 to consider some of the journeys and lives of the great explorers who employed camels in their travels, treating them in order of their dates of birth. They have been selected on the basis of their views on camels and on the significance of their exploration.

CHAPTER TWO

# CAMELS AND THE EXPLORATION HISTORY OF NORTH AFRICA

## INTRODUCTION: THE NORTH AND EAST AFRICAN DESERTS

The Sahara is the world's largest desert, covering an area similar in extent to that of the USA (**Figure 2.1**). Occupying approximately one third of the African continent, it stretches from the Atlantic to the Red Sea and from the Mediterranean to the Sahel zone of West Africa. At its centre, mean annual rainfall drops to less

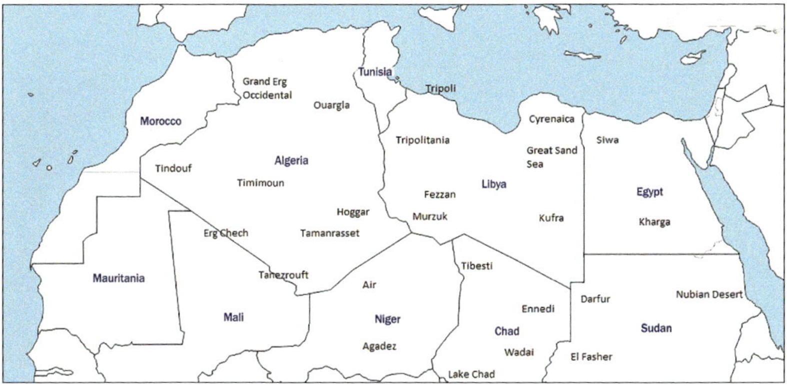


**Figure 2.1** Map of North Africa showing some of the main locations referred to in the text. *Source: Author.*

than 10 mm. Vegetation is largely absent. Sometimes the Sahara is subdivided into two or more parts, with the eastern portion being called the Libyan Desert and the western part being regarded as the Sahara proper.

The Sahara's most distinctive characteristic, save only the relief provided by the Hoggar (over 2900 m above sea-level), Aïr (over 2000 m), Jebel Marra (over 3000 m), Tibesti (over 3400 m) and other massifs, is its flatness. This is associated with great sandstone plateaux, broad, closed basins, wind-moulded landscapes and areas of sand deposition (*ergs*). The Sahara contains some of the world's largest sand seas, and dunes cover just over a quarter of the total area. On its north side, in a belt that runs through the centre of Tunisia westwards towards the high plateaux of Algeria, is a line of depressions – *chotts*. Another great topographic depression on the south side of the Sahara is the Chad Basin. Covering a total area of 2.5 million km$^2$, it stretches for over 1500 km from the low watershed of the Congo into the heart of the Sahara, where it receives runoff from the Tibesti and Hoggar massifs, as well as from rivers entering it from the south. The lowest point of the basin is the dusty Bodélé depression, the altitude of which is only 150 m above sea-level. Lake Chad was visited by such great explorers as Clapperton, Barth, Nachtigal, Foureau, Tilho and Vischer.

The Libyan Desert is bounded in the east by the Nile Valley. It is a huge stretch of territory that extends hundreds of kilometres from the Mediterranean to the Sudan and from the Nile to the mountains of the central Sahara: Ennedi, Erdi, Tibesti, Tummo and Fezzan. In shape and extent it resembles the Indian Peninsula (**Figure 2.2**). For the most part the Libyan Desert is rather flat and only limited areas reach altitudes greater than a few hundred of metres above sea level. Much of it is underlain by relatively gently-dipping sedimentary rocks that create low escarpments and gently sloping plateaux. High land tends only to occur in the south west of the region in the Gilf Kebir, a huge erosional remnant of sandstone attaining heights of more than 1000 m above sea-level, and at the crystalline Gebel Uweinat which rises to over 1900 m.

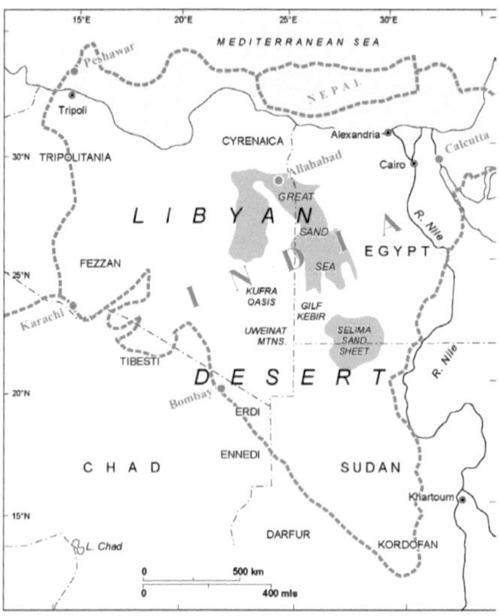


**Figure 2.2** The Libyan Desert – the size of India. *Source: Author.*

Desert winds have moulded fields of dunes – sand seas or *ergs*. The greatest of these tracts is, appropriately, called the Great Sand Sea. Beginning just south of Siwa oasis, it continues almost uninterrupted for 600 km to the Gilf Kebir plateau and covers some 150,000 km$^2$ of eastern Libya and western Egypt. Most of the dunes are great ridges upon which long, parallel ridges, called *seifs*, run more or less north to south in response to the trend of the dominant winds. Because of their height (often more than 100 m), width (often 1 km or more), steepness and instability, they posed a formidable barrier to travel from east to west, from Egypt to Libya. The desert also has some major depressions, often bounded by steep cliffs, which have been excavated in part by millennia of wind erosion. This has also moulded the ground into aerodynamically shaped landforms called yardangs or mud lions, named thus because of their supposed resemblance in shape to those reclining carnivores. The depressions – Fayum, Qattara, Farafra, Bahariya, Dakhla, Kurkur, Kharga and Siwa – are places where underground groundwater reservoirs approach the surface and allow oases of date palms to occur. Some of the depressions are excavated below sea level, most notably Qattara, the base of which lies 133 m lower than the surface of the Mediterranean.

**Figure 2.3** The deserts of north eastern Africa. *Source: Author.*

Between the Nile and the Red Sea lies the Eastern Desert of Egypt, visited by de Bellefonds and Weigall (**Figure 2.3**). In northeastern Sudan and parts of Eritrea lies the Nubian Desert, home of the Bisharin and other nomads, and the Bayuda Desert. Acland (1932) provides an excellent account of the camels of this area. Further south in Eritrea there is the Danakil Desert, home of the Afar people visited by Wilfred Thesiger and Ludovico Nesbitt. Drylands, such as the Chalbi Desert, are also a feature of northern Kenya, and of eastern Ethiopia and Somalia. In northern Uganda, dry conditions exist in Karamoja, and here camels are an important source of livelihoods.

## CHARLES-JACQUES PONCET AND JAMES BRUCE

Charles-Jacques Poncet (born in 1655 in Franche-Comté, died in c.1706 at Isfahan, Persia) was a French resident pharmacist and physician in Cairo who was best known for the account of his travels in Abyssinia (Ethiopia), which had been closed to Europeans after about 1630. Poncet was summoned to Gonder, the Ethiopian capital, to treat Emperor Iyasu I and his son for a skin complaint. His account of the journey, *A Voyage to Ethiopia in the Years 1698, 1699 and 1700* (1949) is the only European source for the history of Ethiopia in the period between the expulsion of the Portuguese in 1633 and the visit of Bruce in 1869–71. He cured the Emperor and his son and made a detailed description of the country during his nine-month stay.

Poncet, who had lived in Egypt since 1687, departed for Ethiopia in May 1698. On his journey he visited Kharga Oasis – the Great Oasis – in the Libyan Desert for four days in 1698. He was the only traveller who left written records of this location between the sixteenth and nineteenth centuries. Accompanied by one Hagi Ali, an officer of the Abyssinian Emperor, and by a Jesuit missionary, Charles Francis Xaverius de Brevedent, Poncet set out from the town of Manfalut in the Nile Valley and crossed to Kharga (**Figure 2.4**). His description of this portion of the journey is given in H. J. Llewellyn Beadnells' *An Egyptian Oasis* (1909) as follows (p.13–15). I have not attempted to modernise the English.

> We set forward on the 2d of October early in the Morning, and from that very Day we enter'd a frightful Desart. These Desarts are extremely dangerous, because the Sands being moving are rais'd by the least Wind which darken the Air, and falling afterwards in Clouds. Passengers are often buried in them, or at least lose the Route, which they ought to keep.

**Figure 2.4** Crescentic dunes in the Kharga depression. *Source; Author.*

Referring to the deserts which surround the oases, Poncet remarks:

> Those vast Wildernesses, where there is neither to be found Bird, nor wild Beast, nor Herbs, no nor so much as a little Fly, and where nothing is to be seen but Mountains of Sand, and the Carcasses, and Bones of Camels, imprint a certain horrour in the Mind, which makes this Voyage very tedious and disagreeable. It wou'd be a hard matter, to Cross those frightful Desarts without the Assistance of Camels. These Animals will continue six or seven Days, without either eating or drinking, which I cou'd never have believ'd, if I had not observ'd it very particularly.

From Kharga, Poncet travelled along the Darb el-Arbain, the well-known caravan route to the south, via Selima, where they watered their camels at 'an excellent spring', Shebb and Dongola. At the end of a journey lasting exactly a year and a day since leaving Cairo, the doctor's caravan crossed the Ethiopian frontier on 11 June, 1699. At the border town of Girana he exchanged his camels for horses. After his sojourn in Ethiopia, Poncet returned via Jedda, Yenbu, Sinai and Cairo.

James Bruce (1730–1794), the giant, red-headed Scottish explorer, also visited Abyssinia in his search for the source of the Nile and returned northwards via Gondar across Sudan to Aswan. He and his team of camels had a hard time (*Travels to Discover the Source of the Nile*, 1790, volume 4, 584–5), beset by heat, shortage of water, and the *simoom* dust storm (**Figure 2.5**). He was forced to jettison his hard-won collections:

> I here began to provide for the worst. I saw the fate of our camels approaching, and that our men grew weak in proportion; our bread, too, began to fail us, altho' we had plenty of camels flesh in its stead; our water, though in all appearance we were to find it more frequently than in the beginning of our journey, was nevertheless brackish, and scarce served the purpose to quench our thirst; and, above all, the dreadful simoom had perfectly exhausted our strength, and brought upon us a degree of cowardice and languor that we struggled with in vain; I therefore, as the last effort, began to throw away every thing weighty I could spare, or that was not absolutely necessary, such as all shells, fossils, minerals, and petrefactions that I could get at, the counter-cases of my quadrant, telescopes, and clock, and several such like things.

Bruce's achievements were not fully appreciated in his lifetime. As John Sabini put it (1981, 41), 'His behaviour was often that of a one-man gunboat'. In 1794 he died through falling downstairs in the haste with which he ran to a staircase to show a lady to her carriage.

**Figure 2.5** The 'dreadful *simoom*' encountered by James Bruce, *Source: Anon.* Pictorial Records of the English in Egypt, *undated.*

# WILLIAM GEORGE BROWNE

William George Browne (1768–1813), born in London, was the son of a wine merchant. He read law at Oriel College, Oxford. Now largely forgotten, he was one of the earliest and most tenacious British explorers of Egypt, Sinai and the Sudan. Left a moderate inheritance by his father, and inspired by the travels of James Bruce in pursuit of the source of the Nile, this thin, grave, well-educated man abandoned his legal studies and determined to become an African explorer. He arrived in Alexandria, Egypt, in 1792. In that year he visited the Siwa Oasis, the first European known to have done so since antiquity. He went to many parts of Egypt, including the Wadi Natrun lakes, Fayum and the oases of the Western Desert. Early in 1793 he went to Sinai.

In May of that year, he set off on his greatest journey to Darfur and Kordofan, reaching Darfur by caravan in July 1793. He travelled along the Darb el-Arba'in via Sheb and Selima. A large proportion of the nearly 500 camels that his caravan had started with died on the way, and some of the merchants had to bury their goods to be collected by them on a future occasion.

The Darb el-Arba'in was noted for its difficulty, especially between Kharga and Assiut. Count Almásy, who used the road in the 1930s, was greatly struck by the Darb el-Arba'in and the evidence it provided in the form of countless skeletons of human and bestial misery (*The Unknown Sahara*, 21). Browne (1799) described his difficult descent with his camels down the great escarpment above Kharga (**Figure 2.6**) (*Travels in Africa, Egypt and Syria, from the Years 1792 to 1798,* 184):

> It forms the Western side of the ridge, which constitutes, as it were, the wall of Egypt, and the Eastern boundary of the low desert, in which lie the *Oases*. It consists of a coarse tufa, and is of rugged and difficult descent. The road seems in many places to have been opened by art. We were a full hour in reaching the bottom. The camels not without great pain carrying their loads on the steep declivity, and being often in danger of falling. From the summit of this rock the view lost itself in an extensive valley, consisting chiefly of rocks and sand, but diversified by small bushes of the date tree, and other marks of vegetation, near the spring where we designed to repose. Nothing could exceed the sterility of the mountain we had passed.

In Darfur, Browne had an exceptionally tough time – he contracted dysentery, was robbed of most of his property and was detained by the Sultan for three

**Figure 2.6** The Refuf Pass down into the Kharga Oasis. The dark masses are tufa. *Source: Author.*

years. He returned to Egypt in 1796, and then visited Syria before returning to London in 1798. People recognised that this book is full of information, as it indeed is, but it was criticised for its abruptness and dryness of style. He has been described as solid rather than shining. The Revd. Robert Walpole, rector of Christ Church, Marylebone, in 1820, remarked that 'it contains some passages offensive to good taste, and a few that are more seriously objectionable'. It is indeed full of material that would not perhaps have been written about in more puritanical Victorian times. He does, for example, discuss venereal diseases, aphrodisiacs, and circumcision. It is also immensely valuable as an early and detailed account of Darfur.

He gave full descriptions of the countryside, the people, and the animals. Of the local camels he wrote (p.258–9):

> There are few countries where the animal abounds more than in Dar-Fur. They are remarkable for enduring thirst, but not for bearing great burthens. The camels in this country are particularly subject to the mange, (*Gcraby*) which attacks them chiefly in winter, and in some pastures much more than in others. This malady is very contagious. It is cured by the application of a kind of tar, procured by distillation per deliquium, from

> the seeds of the water-melon. When the male camel is found unruly, they sometimes deprive him of one or both testicles. It is a cruel operation, as immediately after having incided with an ordinary knife, they sear the wound with an hot iron till the hemorrhage be stopped.

He also discussed the value of their meat and milk:

> The flesh of the camel, particularly of the female, (*naka*), which is fattened for the purpose, is here much used for food. It is insipid, but easily digested, and no way unpleasant. The milk also is in much esteem.

On his return from Africa, Browne, who admired many aspects of oriental life, subsequently travelled to Turkey, Armenia and Persia, where he was murdered by Kurdish bandits in 1813. Prior to that, in London, he met up with the young Jean Louis Burckhardt, and did much to inspire his subsequent travels in Africa and the Middle East. Burckhardt (1822, 347) acknowledged:

> Mr Browne, whose talent and perseverance were such as will seldom be found united in the same person, whose friendship for me I can never forget, and to whose excellent advice I owe much of my success.

## FRIEDRICH HORNEMANN

Friedrich Hornemann (1772–1801?), a clergyman's son, was born in Lower Saxony in Germany. In 1796 he offered his services to the African Association in London as an explorer. The Association was seeking to resolve the mystery of the source of the Niger, the direction of which had been a mystery since ancient times. One tactic was to approach from the west (as Mungo Park had done) and the other was to approach it from the north by crossing the Sahara. This is what was chosen for Hornemann. The Association sent him to the University of Göttingen to study Arabic, and in September 1797 he arrived in Cairo where he continued his studies. A year later he joined a caravan returning to North Africa from Mecca and was accompanied by Joseph Frendenburgh, a German convert to Islam. Masquerading under the name of Yussuf, Hornemann pretended to be a Mamluke (slave warrior). Their destination was Fezzan in southern Libya. They reached Murzuk after two and a half months of travel in November 1798. His companion died there, which Hornemnann ascribed to his being 'led astray by wine and women'.

After setting out from Cairo, Hornemann headed in a caravan to Siwa. He was not the first European to visit Siwa. That prize had been claimed six years earlier by W.G. Browne.

He drew attention to the water skins that his caravan employed (*Travels from Cairo to Mourzouk*, 1802, 4):

> The water collected for the use of the caravans is carried in bags made of goat-skins, unripped in the middle, and stripped from the animal as entire as possible; those made at Soudan are the strongest and best; water may be preserved in them for five days, without acquiring any bad taste: the bags of an inferior manufacture give an ill taste, and a smell of the leather, from the second day. To render the skins flexible and lasting, they are greased on the inside with butter, and by the Arabs sometimes with oil, which latter gives quickly a rancid taste, and to any but an Arab, renders the water scarcely fit for drinking.

He also discussed camel butchery (p.5):

> The slaughtering of a camel affords a feast to the camel drivers and slaves. The friends of the owner of the beast have a preference in the purchase; and after dividing the carcase, every slave comes in for a share: no part of the animal capable of being gnawed by human tooth, is suffered to be lost; the very bones pass through various hands and mouths, before they are thrown away. They make sandals of the skin, and they weave the hair into twine.

He then reached the oasis of Siwa (p.14):

> Siwah is built upon, and round, a mass of rock; in which, according to tradition, the ancient people had only caves for their habitation. Indeed the style of building is such, that the actual houses might be taken for caves; they are raised so close to each other, that many of the streets, even at noon, are dark, and so intricate, that a stranger cannot find his way into or out of the town, small as it is, without a guide . . . The people of our caravan compared Siwah to a bee-hive, and the comparison is suitable, whether regarding the general appearance of the eminence thus covered with buildings, the swarm of its people crowded together, or the confused noise, or hum and buz from its narrow passages and streets, and which reach the ear to a considerable distance.

He described the people (p.18):

> I can give no favourable account of the character of the people of Siwah, either from general repute, or from my own observation. I found them obtrusive and thievish. Our tents, and especially my own, were constantly surrounded and infested by this people; and our merchants were under the necessity of guarding their bales of goods, with more than ordinary attention, under apprehension not merely of pillage, but of general and hostile attack.

Having reached Fezzan, Hornemann tried a local delicacy or two in one of the villages (p.59):

> Here for the first time I was regaled with the great Fezzan dainty of locusts or grasshoppers, and a drink called *lugibi*. The latter is composed of the juice of date trees, and when fresh is sweet and agreeable enough to the taste, but is apt to produce flatulencies and diarrhoea. At first I did not relish the dried locusts, but when accustomed, grew fond of them: when eaten, the legs and wings are broken off and the inner part is scooped out, and what remains has a flavour similar to that of red herrings, but more delicious.

In Murzuk (**Figure 2.7**) Hornemann was not impressed by the local inhabitants (p.70):

> The mein, the walk, and every motion *and* gesture of the people of Fezzan, denote a want of energy, either of mind or body. The tyrannic government, the general poverty of the country, and their only food consisting of dates, or a kind of farinaceous pap, with no meat, and rarely with even a little rancid oil or fat, contribute at once to weakness of frame, and dejection of spirit. Even in those parts, where the race may be supposed to be ameliorated by a mixture with the Arabs, there is no energy of character, no industry.

He also commented on the question of alcohol (p.72), noting that the men of Fezzan were addicted to drunkenness, produced by a date wine called *busa*.

Most importantly, Hornemann discussed the various caravans that made Murzuk such a centre for commerce at that time (p.64–5):

**Figure 2.7** Murzuk in the nineteenth century. Engraved by Barbant. In the *The Universal Geography*, 1878–94, Élisée Reclus, edited by A.H. Keane, published by J.S. Virtue & Co., London.

> The caravans coming to Mourzouk from the south or west, bring, as articles of commerce, slaves of both sexes, ostrich feathers, zibette, tiger skins, and gold, partly in dust, partly in native grains, to be manufactured into rings and other ornaments, for the people of interior Africa. From Bornou copper is imported in great quantity. Cairo sends silks, melayes (striped blue and white calicoes), woollen cloths, glass, imitations of coral, beads for bracelets, and likewise an assortment of East India goods. The merchants of Bengasi, who, usually join the caravan from Cairo at Augila, import tobacco manufactured for chewing, or snuff, and sundry wares fabricated in Turkey. The caravan from Tripoly, chiefly deals in paper, false corals, fire-arms, sabres, knives, and the cloths called abbes, and in red worsted caps. Those trading from Gadames, bring nearly the same articles. The smaller caravans of Tuaricks and Arabs, import butter, oil, fat, and corn; and those coming from the more southern districts bring senna, ostrich feathers, and camels for the slaughter house.

Hornemann remained in Murzuk until June 1799, before going back to Tripoli, from where he dispatched his journals to London. He then returned to Murzuk with the intention of going south to Hausaland. He died, probably in February 1801, but precisely where and how is not known. It was probably on the banks of the Niger. The African Association published his journals as *Travels from Cairo to Mourzouk* (1802). Hornemann was the first European in modern times to have crossed the Sahara from north to south, and the first ever to have done so from east to west.

## HUGH CLAPPERTON

**Figure 2.8** Hugh Clapperton.
*Source: frontispiece in Clapperton 1828.*

Hugh Clapperton (**Figure 2.8**) (1788–1827), the 'Sailor in the Sahara', was a Scottish naval officer who was born in Dumfrieshire. He took part in an early journey across the Sahara from Tripoli to Murzuk and on to Lake Chad. The second member of the expedition was Dixon Denham (1786–1828), a soldier who has been described as odious, pushy, self-important and malicious. He spread unsubstantiated gossip that Clapperton had had sexual relations with an Arab. This charge was a revengeful invention by a man Clapperton had discharged on suspicion of the theft of the carcass of a young camel. It is very plain from Clapperton's diaries that he had a keen interest in ladies of all races.

The other member of the team was Walter Oudney (1790–1824), a pale, grave, slightly built, Scottish naval doctor. He was a neighbour of Clapperton in Edinburgh and aroused his interest in African travel, having himself been intrigued by the life of fellow Scotsman Mungo Park and his exploration of the Niger. The trio were the first Europeans to see Lake Chad. The record of their journey is recorded in Denham 1828. Clapperton's journal, 'a spontaneous, modest and uncomplicated record', has recently been published as *Difficult and Dangerous Roads*, edited by James Bruce-Lockhart and John Wright (2000).

In the early nineteenth century Tripoli was the scene of great British interest. This was partly because it was close to Malta (which had been occupied since 1800) and partly because it was thought to be a good base for exploration into the interior of Africa, and in particular to determine the course of the Niger. The British consul general there, Colonel Warrington Hanmer, strongly encouraged exploration during his long tenure of the post (1814–1846).

Also, Yusuf Pasha, who had usurped the throne of Tripoli in 1795, was keen to project Tripoli's power. So it was that the expedition, the Borno (Bornou) Mission, left Tripoli in early 1822, passing via Garian to Murzuk. The officers went on horseback, accompaniued by 32 camels, a pair of mules and a couple of dogs. En route they were assaulted by a brutal sandstorm. It took eleven months to reach Lake Chad. They often rode on a horse, but also used camels extensively, especially for carrying baggage and water-skins. They did not travel in disguise. The three participants were men of wildly different temperaments and did not get on.

In the introduction, Denham explained the good and bad aspects of the outward desert crossing:

> It will, perhaps, be thought by some, that I have been more minute than necessary in the account of our journey across that tremendous desert which lies between Mourzuk and Bornou, and which, generally speaking, is made up of dark frowning hills of naked rock, or interminable plains, strewed in some places with fragments of stone and pebbles, in others of one vast level surface of sand, and, in others again, the same material rising into immense mounds, altering their form and position according to the strength and direction of the winds. But, even in the midst of this dreary waste, towns, villages, wandering tribes, and kafilas, or caravans, sometimes occur to break the solitude of this dismal belt, which seems to stretch across Northern Africa, and, on many parts of which, not a living creature, even an insect, enlivens the scene.

On New Year's Day 1823, a camel died from exhaustion and a few days later four others collapsed. The Arabs, according to Denham, waited 'in savage impatience, with the knives in their hands, ready on the signal of the owner to plunge them into the animal, and tear off a portion of the flesh for their evening meal' (Hibbert 1982).

They had good things to say about the Arabs they encountered (p.58):

> Arabs are generally thin meagre figures, though possessing expressive and sometimes handsome features, great violence of gesture and muscular action. Irritable and fiery, they are unlike the dwellers in towns and cities: noisy and loud, their common conversational intercourse appears to be a continual strife and quarrel; they are, however, brave, eloquent, and deeply sensible of shame . . . I should, however, without hesitation, pronounce them to be much more cleanly than the lower order of people

> in any European country. Circumcision, and the shaving the hair from the head, and every other part of the body; the frequent ablutions which their religion compels them to perform; all tend to enforce practices of cleanliness.

They gave details of their positive virtues (p.62):

> Arabs have always been commended by the ancients for the fidelity of their attachments, and they are still scrupulously exact to their words, and respectful to their kindred; they have been universally celebrated for their quickness of apprehension and penetration, and the vivacity of their wit. Their language is certainly one of the most ancient in the world; but it has many dialects. The Arabs, however, have their vices and their defects; they are naturally addicted to war, bloodshed, and cruelty; and so malicious as scarcely ever to forget an injury.
>
> An Arab is ever true to his bread and salt; once eat with him, and a knot of friendship is tied which cannot easily be loosened.

Eventually they reached Chad and they remarked upon its beauty (p.182):

> The great lake Tchad, glowing with the golden rays of the sun in its strength, appeared to be within a mile of the spot on which we stood. My heart bounded within me at the prospect, for I believed this lake to be the key to the great object of our search, and I could not refrain from silently imploring Heaven's continued protection, which had enabled us to proceed so far in health and strength even to the accomplishment of our task . . . Flocks of geese and wild ducks, of a most beautiful plumage, were quietly feeding at within half pistol shot of where I stood . . . Pelicans, cranes, four and five feet in height, grey, variegated, and white, were scarcely so many yards from my side, and a bird, between a snipe and a woodcock, resembling both, and larger than either; immense spoonbills of a snowy whiteness, widgeon, teal, yellow-legged plover, and a hundred species of (to me at least) unknown water fowl, were sporting before me; and it was long before I could disturb the tranquillity of the dwellers on these waters by firing a gun . . . The water is sweet and pleasant, and abounds with fish; which the natives have a curious way of catching. Some thirty or forty women go into the lake, with their wrappers brought up between their legs, and tied round their middles, as I should say, by single files, and forming a line at some distance in the water, fronting the land,

> for it is very shallow near the edges, and absolutely charge the fish before them so close, that they are caught by the hand, or leap upon the shore. We purchased some, and the best flavoured was a sort of bream.

After finding and exploring the lake, the party split up. Denham set off to follow the Chari River, while Clapperton and Oudney travelled westwards towards Kano. Oudney, however, died before they reached their destination. Clapperton then went on to Sokoto, before meeting up with Denham back at Kukawa. Together they returned across the Sahara to Tripoli, arriving in January 1825. Clapperton returned to West Africa later in the year, but died near Sokoto on 13 April 1827 as a result of fever and dysentery. Covered with a Union Jack, his body was placed on the back of a camel and taken to the burial pit, where his inconsolable and faithful servant Richard Lander read the funeral service (Lander 1830). Denham died of fever in Sierra Leone, aged 42, in the following year.

The Bornou Mission of 1822–1825 was, in the words of John Wright, 'the most dramatically successful expedition of exploration that the British government had up to then sent into inner Africa.' By the end of the mission, even if the course of the Niger was still unsolved, the central Sahara, Lake Chad, and the central Sudan had all been charted and revealed to the outside world.

## GEORGE FRANCIS LYON

George Lyon (1795–1832) (**Figure 2.9**) was an English naval officer and explorer of North Africa and the Arctic. He was born into a military family in Chichester, and as a young man he joined the Royal Navy. In September 1818, Lyon, serving on HMS *Albion*, was at Malta when Joseph Ritchie (1788?–1819), a Yorkshire surgeon and secretary to the consul in Paris, arrived there on his way to Tripoli to begin his attempt to reach central Africa from the north. Ritchie had been recruited by Sir John Barrow to find the course of the Niger River and the location of Timbuktu. Captain Frederick Marryat, who was to have accompanied Ritchie, proved unable to do so, and Lyon, hoping to advance his professional standing, volunteered to take his place. In November he joined Ritchie at Tripoli. He already had some knowledge of Arabic, and for the next four months studied the language and religious and social customs of the Arabs, adopting the alias 'Said-ben-Abdallah'.

They left Tripoli, together with their servant, Belford, towards the end of March 1819 and reached Murzuk (**Figure 2.10**) on the thirty-ninth day. The

**Figure 2.9** Lyon in a Tripoline outfit. *Source: Lyons 1821.*

expedition headed via Garian towards the Fezzan. In Murzuk they both fell ill, and on 20 November 1819 Ritchie died. Without resources, and still very feeble, Lyon pushed on towards the southern boundary of Fezzan, but he was obliged to return, and reached Tripoli, in very bad shape in March 1820. He had been away for a year. The expedition was written up in *Narrative of Travels in Northern Africa, in the Years 1818, 19, and 20* (1821).

Of the Tuareg he met he was ecstatic (p.110):

> They are the finest race of men I ever saw; tall, straight, and handsome, with a certain air of independence and pride, which is very imposing. They are generally white, that is to say, comparatively so; the dark brown of their complexions only being occasioned by the heat of the climate. Their arms or bodies, (where constantly covered), are as white as those of many Europeans. Their costume is very remarkable, and they cover their faces as high as the eyes, in the manner of women on the sea-coast. Their original motive for so doing is now forgotten; but they say it must be right, as it was the fashion of their forefathers. This covering extends as high as half way up the bridge of the nose, from whence it hangs down

**Figure 2.10** Lyon's depiction of the fourteenth-century castle at Murzuk in 1821. *Source: Lyons 1821.*

below the chin on the breast, much in the same way (but longer) as crape or lace is hung to a lady's half mask. This cloth is generally of blue glazed cotton; but yellow, red, white, and many other colours are worn according to taste, or the ability of the wearer to purchase them. The beard is kept close clipped, so as not to interfere with the covering which is tied behind; their red caps are generally very high, but some wear yellow or green ones, fitted close to the head: others have no caps at all, but leave their hair to grow, and plait it in long tresses. All wear turbans, which are never of any fixed colour: blue is the most common and cheap; but gaudy hues are preferred. A large loose shirt (having the sleeves the same size as the body), called Tobe, is the common dress . . . They all wear a whip, hanging from a belt passed over the left shoulder by the right side. Their swords are straight and of great length, and they wield them with much ease and dexterity. From the left wrist is suspended a dagger, with the hilt towards the hand; it has a broad leather ring attached to the scabbard, and through this the hand is passed. No Tuarick is ever seen without this appendage, and a light elegant spear, sometimes entirely of iron, inlaid with brass; others are of wood, but are also highly ornamented. These weapons are about six feet in length, and are thrown to a great distance.

He was also impressed by their camel riding skills (p.115):

> The manner of riding amongst these people is very singular. They have swift, tall camels, called Maherry (the Herie of travellers), with which they perform extraordinary journeys. The saddle is placed on the withers, and confined by a band under the belly. It is very small, and difficult to sit, which is done by balancing with the feet against the neck of the animal, and holding a tight rein to steady the head. They manage these creatures with great dexterity, fighting when mounted on them, and firing at marks while at full speed, which is a long trot, in which, the maherry can continue at about nine miles an hour for many hours together.

Later in his journey he discussed the matter of drinking the stomach contacts of a camel that had died (p.305):

> I never before had an opportunity of observing how water is procured from the belly of a camel, to satisfy the thirst of an almost perishing Kafflé. It is the false stomach which contains the water and undigested food. This is strained through a cloth, and then drank; and from those who have been under the necessity of making use of this beverage, I learn that the taste is bitter. As this animal had recently drank, its stomach was nearly full. I amused myself in making observations on its skin and skeleton, and in planning the formation of a boat; and I found that a most excellent contrivance might be made from them, for the purpose of crossing rivers; the back-bone being used as the keel, and the ribs as timbers. The formation of the chest of a camel resembles the prow of a Portuguese bean-cod, or fishing-boat. Indeed, it was in consequence of hearing the Arabs always calling it Markab, or ship, that the idea first occurred to me.

His book is engaging, beautifully illustrated and informative. It provides a full early picture of Murzuk, the Fezzan and groups like the Tibbu (Tubu) and the Tuaregs (**Figure 2.11**). Immediately after this trip, he explored the Arctic. He had a genuine interest in the 'natives' of the countries he visited. For example, wearing a thobe and speaking fluent Arabic, he managed to blend in with the inhabitants of the Sahara; he was tattooed by the Inuit in the Arctic, using needle and sooty thread, and ate raw caribou and seal meat with them. He was made an honorary Doctor of Civil Laws (DCL) by the University of Oxford in 1825 and he was elected a Fellow of the Royal Society in 1827. He died on

**Figure 2.11** Lyon's depiction of a Tuareg warrior on a riding camel – 'Maherry'.
*Source: Lyons 1821.*

8 October 1832 on board the ship *Emulous* en route from Buenos Aires to Britain to be treated for eye problems.

## RENÉ CAILLIÉ

René Caillié (1799–1838) was the first European to return alive from Timbuktu, a town on the south side of the Sahara in what is now Mali. Born in Mauzé, France, his background was far from propitious, for his father was a drunkard and a convict, and he himself was thin and frail-looking. Both his parents died while he was still young and he received a limited education. However, the reading of Defoe's *Robinson Crusoe* is said to have whetted his appetite for travel, and in 1816, never having even been to Paris, he set out for Senegal as an officer's servant. The officer soon died, but Caillié decided to remain, and did some gruelling preparatory journeys in the region. It was, however, in 1824 that he decided that he would seek to get to Timbuktu, and that he would wear the clothes of a Moor, behave as a Moor, learn a language the Moors would understand, learn how to be a good Mohammedan, and travel light. He spent considerable time studying the Koran and learning about African Islamic culture. On 3 August, 1824, masquerading in Muslim garb (**Figure 2.12**),

**Figure 2.12** Caillié meditating on the Koran and taking notes. *Source: Caillié 1830, Vol. 2, 72–73.*

Caillié departed for the interior. To that end he spent eight months with the Brakna Moors, who lived north of the Senegal River. He told people that he had been born in Egypt of Arab parents, had been carried away by French soldiers who had invaded Egypt in his infancy, and that his master had bought him to Senegal. His ambition, he said, was to return to Egypt. In 1827, he set out with a Mandingo caravan across Sierra Leone and Guinea to the upper Niger at Kurussa, and on to Djenné, whence he continued to Timbuktu by river. Caillié's impetus for travel into the interior was strengthened when he learned that the Geographical Society of Paris had offered a Gold Medal and a 10,000-franc reward to anyone who could prove that he had been to Timbuktu.

Most of the route he did on foot, but he suffered grievously from fever, an ulcerated foot, and scurvy. He was forced to tarry in the village of Tiémé for some time so that he could recover. He described his feelings when he realised that he had scurvy:

> I soon experienced all the horrors of that dreadful disease. The roof of my mouth became quite bare, a part of the bones exfoliated and fell away, and my teeth seemed ready to drop out of their sockets...To crown my misery, the sore in my foot broke out again and all hope of my departure vanished . . . Alone in the interior of a wild country, stretched on the damp ground, with no pillow but the leather bag which contained my luggage, with no medicine . . . I was soon reduced to a skeleton . . . Suffering had deprived me of all energy. One thought alone absorbed my mind – that of Death. I wished for it, and prayed for it to God.

After recovery, which was aided by the attentions of an old negress who nursed him, he reached Timbuktu in April 1828. This was twelve years after he had left Mauzé and a year after he had started his journey at Kakandé in Senegal. Although Laing had reached Timbuktu before Caillié (in 1826), he had been murdered on leaving the city by the Tuaregs, and so Caillié was the first European to travel there and survive. For this in due course he received the prize of 10,000 francs that had been promised. In 1830 he published his two-volume account of his epic journey in his *Travels through Central Africa to Timbuctoo.* To some people it is a rambling and disappointing account, full of trivial detail, but it is also honest, not least about Timbuktu, and is generally informative about the people he encountered.

Timbuktu proved to be a great disappointment to Caillié, for it was smaller than he had expected, with, he estimated, only 10,000–12,000 inhabitants. Its commerce appeared to be limited, it had poor water, it was dreary, the market was inactive, and the area was terrorised by the Tuaregs.

On the other hand he liked the people, and especially the women (Caillié 1830, Volume 2, 61):

> The women of Timbuctoo are not veiled like those of Morocco: they are allowed to go out when they please, and are at liberty to see any one. The people are gentle and complaisant to strangers. In trade they are industrious and intelligent; and the traders are generally wealthy and have many slaves. The men are of the ordinary size, well made, upright, and walk with a firm step. Their colour is a fine deep black. Their noses are a little more aquiline than those of the Mandingoes, and like them they have thin lips and large eyes. I saw some women who might be considered pretty.

Caillié noted that for the Tuaregs, the largest and fattest women were the most admired, especially those who were so obese that they were unable to walk without being supported by two assistants.

He was concerned about the possibility of attacks by the Tuareg (p.65):

> These people visit Timbuctoo for the sole purpose of extorting from the inhabitants what they call presents, but what might be more properly called forced contributions . . . When the chief of the Tooariks arrives with his suite at Timbuctoo, it is a general calamity, and yet every one overwhelms him with attention, and sends presents to him and his followers. He sometimes remains there two months, being maintained all that time at the expense of the inhabitants and the king, who sometimes give them really valuable presents, and they return home laden with millet, rice,

> honey, and preserved articles . . . The Tooariks have terrified the negroes of their neighbourhood into subjection, and they inflict upon them the most cruel depredations and exactions. Like the Arabs, they have fine horses which facilitate their marauding expeditions.

Caillié stayed in Timbuktu (**Figure 2.13**) for two weeks and then joined a caravan of some 600 camels that was going north to Morocco. This journey was one which no European was known to have attempted. He almost missed the caravan as it departed. He had to run a mile through sand to reach it. He complained (p. 88):

> This effort fatigued me so much, that, on reaching the caravan, I fell down in a state of insensibility; I was lifted up and placed on a loaded camel, where I sat among the packages, and though dreadfully shaken I was too glad at being relieved from the labour of walking to complain of my beast.

**Figure 2.13** Timbuktu as portrayed by Caillié. *Source: Caillié 1830, Vol. 2, 71–73.*

From El Arawan he was with an even bigger caravan of 1400 camels, laden with 'the productions of the Soudan': gold, slaves, ivory, gum, ostrich-feathers, and cloth. It took in all about three months to reach Fez and he was subjected to times of great thirst, to sand storms, to mirages, to insults concerning the colour of his skin, to scorpions and to snakes. He was also injured when he fell off a bolting camel near El-Harib, but was helped by a stranger, a Moor, who helped him remount.

He was graphic in his account of a sandstorm (p.103):

> On the 14th of May, a violent gale blew from the east, which unroofed several of the houses, and raised such a quantity of sand, that it was impossible to keep the doors open. The heat, though there was no sun, was stifling. The air was full of sand, which descended in the night. It would be impossible to express what I suffered during this storm. I was obliged to lie on the ground, with my head enveloped in a pagne, to protect myself from the burning sand, which entered through the chinks in the door. I experienced a continual thirst, and had nothing but warm and brackish water to quench it. This unwholesome drink caused a violent derangement of the stomach, and the heat, rising to a degree I had never before felt, produced a dreadful head-ache.

Mirages were also a frustration (p.125):

> In crossing the desert, I perceived, at a distance, immense tracts, which had the appearance of rivers or lakes, with islands of sand rising in the midst of them; they presented themselves to the eye, in the horizon of the desert, as places where one might quench one's thirst. This prospect broke for a time the uniformity of these vast deserts; on approaching, I was cruelly disappointed, for the water vanished, and I saw nothing but loose sand where I had hoped to quench my thirst.

It was a great relief to him as they reached the oases of southern Morocco. Back in France, Caillié received the prize offered by the Paris Geographical Society, was awarded a pension, and bought a farm near his birth place of Mauzé. He got married and sired four children in quick succession. A man who exuded goodwill and stoicism, he died in 1838, aged 38, from consumption and a recurrent fever caught in Africa. His account of Timbuktu encouraged later explorers, including Heinrich Barth.

# HEINRICH BARTH

Heinrich Barth (1821–1865) (**Figure 2.14**) was born in Hamburg and educated at the University of Berlin, where he was a student of the esteemed geographer Karl Ritter. He was one of the most notable Saharan explorers. Unlike many of the great explorers he was essentially a scholar and linguist, who had been stimulated by such great German geographers as Alexander von Humboldt and Karl Ritter. He spoke and wrote English, French and Italian fluently, and he was also an authority on Arab culture and language. Barth's expertise in archaeology, ethnology and linguistics is vividly displayed often in all of his writings. Moreover, he did not return from his travels with a slim journal merely cataloguing his route, events and a few meteorological statistics. He produced a large five volume account, which, though somewhat dull, is a masterpiece of geographical, historical, anthropological and philological research. He was also notable for his generally sympathetic attitude to the people through whose lands he travelled. He had great strength and physical stature, being well over six feet (1.8 m) tall. He was also stubborn, thorough, did not suffer fools gladly, and was humourless, but mentally tough, which he needed to be, for his greatest journey lasted five years and five months. It involved a distance of 16,000–19000 km.

After studying Arabic in London, he embarked on his first African journey in 1845, heading from Tangier to Egypt, up the Nile to Wadi Halfa, and then to Sinai, Syria, Turkey and Greece. In 1849 Barth joined the quaintly named 'Mixed Scientific and Commercial expedition to Central Africa'. At the request of Lord Palmerston it was commanded by Captain James Richardson (1806–1851), a fervent anti-slavery campaigner who had already crossed the Sahara to Ghat, and also involved a noted German geologist and astronomer, Adolf Overweg (1822–1852). The three men left Tripoli in the spring of 1850.

**Figure 2.14** Heinrich Barth.
*Source: Bibliothèque nationale de France.*

**Figure 2.15** The road southwards from Tripoli to Fezzan at Garian. *Source: Author.*

As he said in *Travels and Discoveries in North and Central Africa* (1890, 40) he was fortunate to have an Arab camel which was his faithful companion as far as Kukawa in Bornu. Leaving Tripoli they passed through the great Garian Escarpment (**Figure 2.15**).

They then went through Murzuq and reached the Tuareg town of Tin Tellust in the autumn. They experienced the Tassili-n-Ajjer and Aïr Mountains and Barth made a detour to the decayed city of Agadez. He entered that city with a stained skin, and wearing a black tobe and a white burnous, but his disguise was not convincing.

In January 1851, they reached Damergou, which lies on the south side of the Sahara proper. At this point Barth and Overweg parted ways from Richardson (an unpredictable and autocratic leader), but planned to reunite with him on the shores of Lake Chad. In the event this was not possible, for Richardson was shortly to die from fever and exhaustion in March 1851. Barth and Overweg also split up, with Barth heading off towards Kano, a city in what is now northern Nigeria. Overweg also died from malaria in September 1852 near Lake Chad.

From Kano and Kukawa, Barth travelled to the Niger and reached Timbuktu in September 1853, having compared the people of Kano and Bornou (Borno) (Barth 1890, 385):

> There is a great difference of character between these two towns . . . the Bornu people are by temperament far more phlegmatic than those of Kano. The women in general are much more ugly, with square short figures, large heads, and broad noses, with immense nostrils, disfigured still more by the enormity of a red bead or coral worn in the nostril. Nevertheless they are certainly quite as coquettish, and, as far as I had occasion to observe, at least as wanton also, as the more cheerful and sprightly Hausa women.

To reach Timbuktu, Barth assumed the guise of a travelling Muslim by the name of Abd el Kerim. He stayed there until March the following year, before returning to Kukawa (Kuka). He then went back across the Sahara to Tripoli, travelling in the heat of summer, and reaching there in August 1855. Eduard Vogel (1829–1856), sent later to central Africa to help Barth, also died, beaten to death with an iron-tipped cudgel in Wardai. Moritz von Beurmann, who went in search of Vogel, was also killed in Wardai. Barth was a survivor.

Barth, quoted in Marozzi (2010, 201), showed an appreciation of the much vilified camel:

> I am rather afraid that Europeans often make camels stupid by their own foolish treatment of them, whereas I was wont to treat this noble animal, which had carried myself or the heaviest of my things all the way from Tripoli, as a sensible companion, giving it in the beginning the peel of the oranges I was eating, of which it was particularly fond, or a few of my dates (for which it did not fail to turn round its beautiful neck) . . . Its fidelity will ever remain in my memory as one of the pleasantest recollections of my journey.

Prothero (1958) summarised Barth's significance thus:

> He had been away for nearly five and a half years and during that time had travelled a total distance of over 10,000 miles. His single most important discovery was of the upper reaches of the River Benue; second to this were his journeys between Sokoto and Timbuktu. In the course of the latter he had spent a longer time in Timbuktu than René Caillié, the only European previously to visit the city and return alive. On his return journey from Timbuktu to Say along the Niger valley he was passing through country seen previously by Mungo Park,

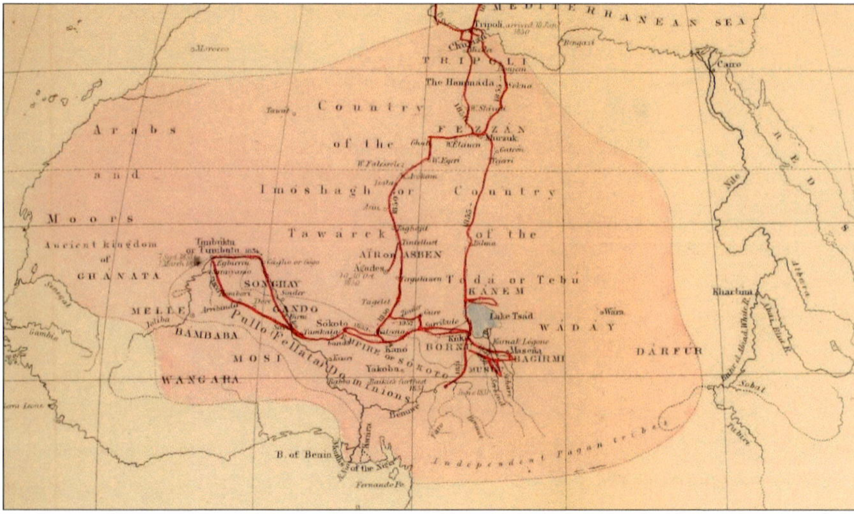


**Figure 2.16** Barth's route, 1850–1855, from his 1859 book.

> whose death further downstream had prevented any report of it from reaching the outside world. Between Kukawa and Sokoto the country had been traversed by Clapperton and in part by Richard Lander, but knowledge of it was advanced greatly by Barth's observations. Even though he made no startling original discoveries the range of his journeys was immense. What is most important is that he recorded in great detail all that he saw and heard. The immense industry displayed in his journals is amazing.

His five-volume book, *Travels and Discoveries in North and Central Africa* (1857–1858), comprising more than 3000 pages, was published simultaneously in English and German. He had wandered through, described and mapped huge portions of the Sahara, Lake Chad-Benue River region and segments of the Niger River (**Figure 2.16**); he had provided for the first time reliable ethnological and linguistic accounts of African tribes; and he was the only European to enter Timbuktu between René Caillié in 1828 and Oskar Lenz in 1880.

The RGS awarded him its Patron's Medal in 1856, Oxford gave him an Honorary Degree, but he died in Berlin of a stomach disorder, probably a result of his great journey, in 1865, aged just 44.

## SIR SAMUEL AND LADY BAKER

Sir Samuel Baker (1821–1893) and his wife were two of the most important travellers who sought the source of the Nile in the middle of the nineteenth century. Baker was born in London, the offspring of a wealthy commercial family. His father was a sugar merchant, banker and ship owner. His second wife, Florence, who accompanied him on his journeys, had an extraordinary history. Hailing from Transylvania, Sam had met her at a slave market in Turkey, where she was in the process of being sold. He acquired her and then fell in love with her.

In March 1861, they started upon their first tour of exploration in central Africa. They arrived in Khartoum, which they left in December 1862 to follow up the course of the White Nile. Two months later at Gondokoro (Juba) he met Speke and Grant. They then pressed on towards the border of what is now Uganda, passing through Latooka country, and eventually achieved the first European visit to Albert Nyanza (Lake Albert). Baker first sighted the lake on 14 March 1864. Baker wrote up the journey in his book, *The Albert Nyanza Great Basin of the Nile and Exploration of Nile Sources* (1866).

Baker was impressed by the physique of the Latooka men, who, with the exception of their bracelets, necklaces and earrings, were generally totally naked: (p.129):

> The Latookas are the finest savages I have ever seen . . . Not only are they tall, but they possess a wonderful muscular development, having beautifully proportioned legs and arms: and although extremely powerful, they are never fleshy or corpulent . . . They all have a remarkably pleasing cast of countenance, and are a great contrast to other tribes in civility of manner.

He had firm views about Africans (p.183):

> From a long experience with African savages, I think it is as absurd to condemn the negro in toto, as it is preposterous to compare his intellectual capacity with that of the white man.
>
> The negro has been, and still is, thoroughly misunderstood. However severely we may condemn the horrible system of slavery, the results of emancipation have proved that the negro does not appreciate the blessings of freedom, nor does he show the slightest feeling of gratitude to the hand that broke the rivets of his fetters.

More importantly, Baker had a great deal to say about the camels that had carried their baggage all the way through the Sudan (p.159 et seq.). First of all he wrote of their feeding habits:

> Our conversation was suddenly terminated by one of my men running in to the tent with the bad news that one of the camels had dropped down and was dying. The report was too true. He was poisoned by a well-known plant that he had been caught in the act of eating. In a few hours he died. There is no more stupid animal than the camel. Nature has implanted in most animals an instinctive knowledge of the plants suitable for food, and they generally avoid those that are poisonous: but the camel will eat indiscriminately anything that is green; and if in a country where the plant exists that is well known by the Arabs as the 'camel poison,' watchers must always accompany the animals while grazing. The most fatal plant is a creeper, very succulent, and so beautifully green that its dense foliage is most attractive to the stupid victim. The stomach of the camel is very subject to inflammation, which is rapidly fatal. I have frequently seen them, after several days of sharp desert marching, arrive in good pasture, and die, within a few hours, of inflammation caused by repletion. It is extraordinary how they can exist upon the driest and apparently most innutritious food. When other animals are starving, the camel manages to pick up a subsistence, eating the ends of barren, leafless twigs, the dried sticks of certain shrubs, and the tough dry paper-like substance of the dome palm, about as succulent a breakfast as would be a green umbrella and a Times newspaper.

Baker then turned to their disposition:

> The camel is by no means well understood in Europe. Far from being the docile and patient animal generally described, it is quite the reverse, and the males are frequently dangerous. They are exceedingly perverse; and are, as before described, excessively stupid. For the great deserts they are wonderfully adapted, and without them it would be impossible to cross certain tracts of country for want of water.

He also discussed their drinking habits (p.160):

> Exaggerated accounts have been written respecting the length of time that a camel can travel without drinking. The period that the animal can subsist without suffering from thirst depends entirely upon the season

> and the quality of food. Precisely as in Europe sheep require but little water when fed upon turnips, so does the camel exist almost without drinking during the rainy season when pastured upon succulent and dewy herbage. During the hottest season, when green herbage ceases to exist in the countries inhabited by camels, they are led to water every alternate day, thus they are supposed to drink once in forty-eight hours; but when upon the march across deserts, where no water exists, they are expected to carry a load of from five to six hundred pounds, and to march twenty-five miles per day, for three days, without drinking, but to be watered on the fourth day. Thus a camel should drink the evening before the start, and he will carry his load one hundred miles without the necessity of drinking; not, however, without suffering from thirst. On the third day's march, during the hot simoom, the camel should drink if possible; but he can endure the fourth day.

He praised the qualities of a good camel (p.161):

> The average value of a baggage camel among the Soudan Arabs is fifteen dollars, but a good 'hygeen,' or riding dromedary, is worth from fifty to a hundred and fifty dollars, according to his capabilities. A thoroughly good hygeen is supposed to travel fifty miles a day, and to continue this pace for five days, carrying only his rider and a small water- skin or girba. His action should be so easy that his long ambling trot should produce that peculiar movement adopted by a nurse when hushing a child to sleep upon her knee. This movement is delightful, and the quick elastic step of a first-class animal imparts an invigorating spirit to the rider; and were it not for the intensity of the sun, he would willingly ride for ever. The difference of action and of comfort to the rider between a common camel and a high class hygeen is equal to that between a thoroughbred and a heavy dray-horse.

The Bakers started upon their return journey and reached Khartoum in May 1865. In that October they returned to England. In recognition of his achievements, the RGS awarded Samuel its gold medal, and a similar distinction was bestowed on him by the Paris Geographical Society. In August 1866 he was knighted by Queen Victoria. In 1871–1873 he went on a journey in connection with the suppression of the slave trade in the Sudan. This was published as *Ismailia* (1874). Once again the Bakers employed camels to carry baggage, including boat parts (**Figure 2.17**).

**Figure 2.17** The Bakers' caravan transporting boat parts through Nubia.
*Source: Baker's* Ismailia *1874, 12.*

His obituary in *Nature* (Anon 1894) recounts his final years:

> In 1869 he commenced the occupation of the upper White Nile provinces for the Egyptian Government, at the head of a body of Egyptian troops, and for five years laboured at the heavy task of restraining the slave-dealing Arabs and keeping in order his apathetic and often disaffected Egyptian subordinates. He established steam navigation on the Nile to the equator, and in his 'Ismailia,' published in 1874, told the story of the extension of Egypt. This completed his career as a pioneer and explorer; but a traveller he remained to the very end of life, and . . . he spent almost every winter either in Egypt or in India.

His last book was on wild animals, for Baker was an avid sportsman, and during his remarkable life had slaughtered wild animals on a massive scale. His health kept up to within a month of his death, and to the last he remained a keen hunter. He died in his residence at Sandford Orleigh, Newton Abbot, in Devon. The faithful Florence died in Devon more than twenty years later.

Two other major figures in the search for the source of the Nile were Speke and Grant. Their paths crossed with those of the Bakers. Grant wrote an account of their return journey through the Nubian Desert (Grant 1884). They travelled by camel from Abu Hamed to Korosko (which lies just south of

Aswan). Grant was keen on the scenery but scornful of the settlements which they passed through (p.326):

> Of all the journeys I have made in Africa, Abyssinia included, and in India, from the Khyber Pass to Calcutta and Bombay, this from Abu Hamed to Korosko is the very worst, from its barrenness, its heat, and from the fatigue and discomfort it necessitates. Abu Hamed, at one end, is a miserable hole with its fort and huts almost smothered in sand; and the scattered village of Korosko at the north end of the route is well described by the governor, who was there at the time, and called it as hot as – , for it is the focus of heat, being completely surrounded by steep-sided fiery mountains, which inclose the small confined space to the very huts, no exit except the Nile being visible. Between these two miserable places there is a desert of 230 miles without one drop of water, except once at the Morad wells – not enough grass to fill your hat, not a hut nor a bundle of fire wood in the entire route, nothing but eternal sand to walk over, varied by climbing over the most rugged ridges of rock.

Korosko and Kordofan had also been visited on camel by Captain William Peel, RN, VC (a close relative of Robert Peel, the prime minister), and Joseph Churi (a Maronite from Lebanon) in 1851–1852 (Peel 1852). They suffered from great ill health.

## GUSTAV NACHTIGAL

Gustav Nachtigal (1834–1885), one of the greatest of the German explorers of the deserts of northern Africa, was born in Eichstadt, Saxony,. The son of a Lutheran pastor, he trained as a doctor and became a military surgeon. However, the climate of Germany was not good for his weak chest, and so he went to North Africa in 1862. After a short spell in Algeria, he practiced in Tunis as physician to the Bey. His health was transformed. In 1869 his life was transformed still further when another great German explorer, Gerhard Rohlfs, acting on behalf of King William of Prussia, charged him with conveying gifts to Shehu Umar, Sultan of Bornu, in recognition of the ruler's protection and support for German explorers, including Rohlfs, Barth and Overweg, in the area of Lake Chad. The gifts included a velvet chair of state, portraits of King William and his family, heavy guns and ammunition, Bibles in Arabic, a harmonium, a pendulum clock and a silver tea-service.

Nachtigal set out from Tripoli in 1869, with six men and eight camels. He was aware that camels from the coastal plain near Tripoli might not be fit to face the rigours of the climate in the interior. Conversely, he was exercised by the fact that camels from the interior had found the cold winter season in Tripoli to be debilitating. They had to be specially fed with a diet of nourishing barley. He was accompanied by Muhammad el-Qatruni, who had served both Barth and Rohlfs, and by an Italian baker and chef, Giuseppe Valpreda. Among those who saw them off was Gerhard Rolfs. As Nachtigal mounted his camel he mused as to whether he would ever see family and friends again.

Gustav travelled with compass in one hand and watch in the other, measured and counted paces, and thus reckoned the distances traversed. He also carried a thermometer, an aneroid barometer and a hypsometer so that he could make meteorological observations. He calculated that his camels travelled at around three and a half kilometres per hour. He also noted that there was a weekly postal service between Tripoli and Murzuk, which was undertaken by relays of swift camels, and lasted eighteen days.

He got to Murzuk, deep in southern Libya in five weeks, and there met up with Alexandrine Tinne (1835–1869).

As there was no merchant caravan due to head southwards towards Bornu, Nachtigal decided to make a detour to Tibesti, the country of the Tubu. He was the first European to make a detailed study of this region. His journey was tough, with little water available and he expressed the horrors for him and his camels (p.241):

> We were afflicted by a terrible thirst; the mucous membranes of mouth, throat, nose and larynx lost their last trace of moisture. An iron ring seemed to be pressing closer and closer over the temples and forehead … Twice, despite my blows, my exhausted animal laid his weary limbs down under a tree, and twice, by redoubling the slapping, I succeeded in rousing the poor creature to totter painfully forward. Then when in the middle of the morning for the third time she forced her way into the branches of an acacia whose long tough thorns tore my skin, and lay down there, all the stubbornness of her species was displayed, and nothing could move her to surrender the shade thus painfully won.

He found the Tubu (p.219) to be 'in the highest degree tedious, presumptuous, arrogant beggars'. They imprisoned him for a while, but eventually he was able to escape from Bardai and return to Murzuk.

He gave a graphic account of the Tibesti landscape (Nachtigal 1974, 235), which filled him:

> . . . with an emotion of awe-struck horror, such as children might feel in a graveyard at night. On the vertical walls of the ravines there frequently appeared in the depths grey, red, white, violet, brown or yellow limestone under the colossal covering of dark sandstone. Here were rounded hills covered with immense blocks of stone, there gigantic cubes strewn one above the other, forming either larger horizontal stone surfaces, or, if the ground around were broken or crumbling, real colossi of columns and pillars.

He reckoned that the summit of Tibesti lay at about 2400–2700 m above sea level, and recorded the presence of a crater at the summit.

Eventually he was able to continue to Bornu, where he delivered his gifts. He learned the Bornu language. He then made extensive travels around Lake Chad, in Kanem, and in Bodélé, the huge Saharan depression to the north of Lake Chad. Chad and Bodélé were the prime focus of a paper he delivered to the RGS (*Journal of the Royal Geographical Society*, 1876). He noted the network of papyrus and reeds that often obscured the view of the Lake Chad, its richness in fish and hippopotami, and the nature of the rivers such as the Logon and the Shari that flowed into it. He correctly identified that Chad currently had no outlet, but he suspected that there might have been one at no very ancient date and that Chad and Bodélé had been united, then containing a vast body of water. Subsequent investigations have shown the truth in this.

He also went across to Wadai, where he became the first European to survive a visit. He approached the capital, Abeshr or Abéché, with some trepidation, for his predecessor Eduard Vogel had been killed here only 17 years earlier. Moritz von Beurmann, who had been sent out to learn Vogel's fate, had also been murdered. The people, Nachtigal remarked (1974, 409) 'even surpass the people of Bornu in their hatred of strangers, in their rudeness, and general lack of civilisation'. He then went on to Darfur, including the Jebel Marra, which previously had only been visited by two Europeans, and Kordofan. He reached Cairo on 22 November 1874, and returned to great acclaim in Europe.

His journey had lasted five and a half years, covered almost 10,000 km, and took him to parts where no European had yet been, including the heart of Tibesti. His three-volume account of his travels, *Sahara und Sudan*, makes colourful and riveting reading. Volume 1 (published in 1879, and translated into English in 1974) recounts the first part of the journey, from Tripoli, through Fezzan, and Tibesti to Bornu. Volume 2 (published in 1881), covers the second part of the journey, from Bornu to the sultanate of Baguirmi (present-day Chad) and to Timbuktu (present-day Mali). Volume 3, published in 1889, four years after Nachtigal's death, recounts the final part of the expedition, from Wadai, through Darfur, and on to the Nile at Khartoum. These volumes are highly informative rather than being a mere travelogue or catalogue, for they also contain a great deal of historical, political, topographic, meteorological, natural history and ethnographic data. In contrast to many of his contemporaries, and in spite of some of his experiences with the people of Tibesti and Wadai, he did not seem to believe in the inferiority of Africans, but hoped that European colonisation might stop slave hunting and slave keeping. Indeed, much of his success as a traveller was because of his ability to make friends and to establish a rapport with his African hosts.

Honours were deservedly showered upon him. The Paris Geographical Society awarded him its great gold medal, and in his absence he was elected President of the Berlin Geographical Society, a position which he held until 1882. In that year the RGS awarded Nachtigal its Founder's Medal. Bismarck appointed him as Special Commissioner for West Africa, and in this capacity he was involved with the acquisition of Togoland and Kamerun as the first German colonial possessions. However, he died of fever onboard a German gunboat, the *Möwe* (Seagull), off the west coast of Africa, aged 51. His record was outstanding: He was the first European to have explored the Tibesti area in the central Sahara, the area of Lake Fittri, Kanem and Borku to the N.E. of Chad, and to have emerged through Wadai to Darfur and Khartoum.

## FRIEDRICH GERHARD ROHLFS

Gerhard Rohlfs (1831–1896) (**Figure 2.18**) was a German who explored both the western and eastern portions of the Sahara. John Wright (2006, 310) regarded him as 'The greatest European traveller of the central Sahara'. Rohlfs was a gritty, tough, peripatetic adventurer. He was a scholarly-looking man with a carefully waxed moustache and nicely brushed goatee. Born near Bremen, he went to Algeria in 1855 and enlisted in the Foreign Legion. He took part in the conquest of Kabylia, was decorated for bravery as Chevalier of the Legion of Honour, mastered Arabic, and became familiar with native customs. In 1862 he went to Morocco, presenting himself as a Muslim. With German thoroughness,

**Figure 2.18** Rohlfs dressed in Mohammedan attire. *Source:* Illustrirter Kalender, *Jg. 20 (1865), S. 95.*

he even had himself circumcised. Disguised as a Mohammedan physician, and calling himself Mustafa, he travelled widely in Morocco, traversed the Wadi Draa and reached the Saharan oases of Tafilet and Tuat (which are now in Algeria). On one occasion, while near Tafilet, he was almost killed by marauders who shot him through the thigh, and almost severed his left arm and right hand. This is all described in his book *Adventures in Morocco and Journey along the Oases of Draa and Tafilet* (1874), which is a remarkably full and informative account of the people, climate, politics, religions and diseases (syphilis appears to have been very prevalent), besides giving the narrative of his travels.

In his *Adventures in Morocco* (p.12) he described how he travelled:

> I had reduced my baggage to the merest necessaries, namely a bundle of linen, which I carried on a stick hanging from my shoulder. My dress consisted of a *djelaba*, a long woollen shirt with a hood, yellow slippers on my bare feet, and a Spanish cap, within which I had stitched my whole stock of money, an English five-pound note; finally a black loose English overcoat served as my burnoose; I had no weapons; a small note- book with a lead pencil was hidden in my pocket.

He described the Berbers and the Arabs (p.41):

> The physical type is both the same: slender, sinewy form, with good muscular development, brown skins, Caucasian form of countenance, aquiline noses, dark fiery eyes, smooth black hair, prominent cheek-bones, scanty beards.

He had some remarks about the women he encountered (p.42):

> The Arab women appear to be somewhat shorter in stature than the women of the Berbers; but that is the only difference between them. Both pass prematurely from childhood to youth, have beautiful figures, and regularly-shaped features when they are young, but soon grown old and hideous hags, becoming dreadfully thin, on account of their meagre food, so that their skin hangs about them in folds.

While in the Draa oasis, Rohlfs was nearly murdered (p.361):

> . . . when I awoke I found the Schich of the oasis, my friendly host, standing over me, with the smoking mouth of his long gun still pointing at my breast. Luckily, he had not, as he intended struck my heart, but had only broken my left arm above the elbow. I was seizing my pistol, when he slashed my hand nearly off with his sabre. From that moment, what with pain and loss of blood, which was streaming from my arm, I became unconscious . . . When I regained consciousness next morning, I found myself alone with nine wounds; for, after I had fainted, these ruffians had shot and slashed me, to make sure of me as they thought. They had robbed me of everything but the bloody clothes I had on.

Luckily some locals came to his rescue and he survived.

He also undertook a further series of journeys in Algeria and Tripolitania and crossed the Sahara from Murzuk to Bornu and the coast near Lagos. In 1869 he travelled from Tripoli to Siwa and Alexandria in northern Egypt. His greatest explorations occurred in 1873/4, when the Khedive Ismail commissioned him to travel eastwards across the Libyan Desert. He took 100 camels and 90 men. The expedition commenced from the Nile Valley in December 1873. Rohlfs's original destination was the isolated Kufra Oasis, the home of the Senussi tribe. Rohlfs was told that the region had not experienced precipitation during the past ten years. Realizing that Kufra was impossible to reach, the expedition turned north toward the Siwa Oasis. The expedition arrived there in February 1874. In 1878, on his last expedition, Rohlfs was still determined to see the Kufra Oasis. After ten years of questing, he finally reached it in late 1878, but his stay was short-lived. The Senussi proved to be hostile, and he was driven away rather quickly, so that he had to retreat to Benghazi.

His expedition across the Libyan Desert, written up as *Drei Monate in der Libyschen Wüste* (1875), was notable for the impressive team of scientists that he recruited: Professor Ascherson, a botanist from the University of Berlin; Professor Little, of the University of Munich, a geologist and palaeontologist; Professor Jordan, a geodesist and astronomer from Karlsruhe, and G. Remelé, a photographer from Prussia. The expedition ascended the Nile to Asyut in a steamer provided by the Khedive. From here, with 140 camels, Rohlfs and his colleagues headed west and after eleven days reached the oasis of Farafra, located in the White Desert. They went on to Dakhla before crossing the 100-m high dunes of the Great Sand Sea (**Figure 2.19**) and then on to Siwa, the oasis of Jupiter Ammon. This last phase took fifteen days. They had no guide but the compass, sextant and theodolites they carried, and they encountered no source of water. They returned to Cairo via Dakhla, Kharga, and the Nile. Such impressive scientific exploration of the Libyan Desert would not occur again until the late 1920s, when such people as Ralph Bagnold appeared on the scene in their Ford cars.

Sometimes regarded as the German version of Sir Richard Burton, Rohlf's work was appreciated by the RGS and in 1868 he received its Patron's Medal. He died near Bonn in 1896. His obituary in the *Geographical Journal* (Ravenstein 1896) summed up his career thus:

**Figure 2.19** The Great Sand Sea. *Source: Author.*

> Dr Rohlfs, in the course of his extensive travels, made known to us wide regions in Africa, of which, up to his time, we only knew from the reports of earlier Arab explorers or from native information. He never made any astronomical observations, nor seems his knowledge of natural history to have been very profound; but he furnished excellent accounts of the countries he traversed, and their inhabitants, laid down his routes from compass bearings, and kept a careful record of meteorological observations.

He was also the first European known to have traversed North Africa by land from the Mediterranean Sea south to the Gulf of Guinea and from Morocco to the Nile. Much of this travel was done by camel.

## ROBERT MURDOCH SMITH AND EDWIN PORCHER

The magnificent classical sites of Libya have for long attracted attention. For example, in 1817, an Italian, Paolo Della Cella, went with the Bey of Tunis's army from Tripoli to the borders of Egypt, and described Leptis Magna and Cyrene. Cyrene was also visited by a small camel caravan employed by the Frenchman, Jean-Raimond Pacho (see Pacho 1827). Travelling westwards from Alexandria, he described the ruins of Marmarica and Cyrenaica and copied inscriptions.

**Figure 2.20** Cyrene in 1969. *Source: Author.*

**Figure 2.21** One of Porcher's drawings at Cyrene. *Source: Smith and Porcher 1864.*

He committed suicide quite shortly after his return. Some decades later he was followed into Cyrenaica by two British explorers, Smith and Porcher.

Robert Murdoch Smith (1835–1900) and Edwin Porcher (1824–1876) were two British military men who, having met up in Malta, undertook a camel-based journey from Benghazi to Cyrene in northern Libya in order to carry out archaeological investigations at one of North Africa's most important classical sites (**Figure 2.20**). Cyrene was also the place from which the plant *Silphium* had come.

Smith was a Royal Engineer from Scotland, who self-funded the expedition to Cyrenaica in 1860–61 (Dickson 1901). He was accompanied by Porcher, who was a naval officer and an expert draughtsman and artist. He had already poroduced paintings of Carthage for Nathan Davis. Smith and Porcher were among the first people to use photography to record excavations (Thorn 2007). They employed camels to carry their baggage, and once they reached Cyrene the camels were also used to pull sledges on which the finds were placed. The party consisted of 17 men, 10 camels and 7 horses and the expedition took twelve days. They liked the scenery through which they passed, and described it as looking like English parkland. They arrived in Cyrene in the rain on 23 December, 1860, and stayed for ten months. They made many remarkable finds, which were painted by Edwin (**Figure 2.21**), including a statue of Apollo. Some of these were lodged with the British Museum. They took the opportunity while based at Cyrene to make journeys to other sites, including Appolonia. Their journey and finds were written up in 1864.

After the Cyrene trip, in 1865 Porcher was put in command of HMS *Sparrowhawk*, stationed in British Columbia, where he remained until 1868. After 25 years of naval service, he retired in 1872 and toured the continent until his death in Germany in 1878. Robert went on to a very distinguished career. He became a Major General and spent many years in Persia, before coming the first director of the new Edinburgh Museum of Science and Art.

## ALEXINE TINNE

Alexandrine Petronella Francisca Tinne (1835–1869), who travelled in Egypt, Sudan, Algeria and Libya, was also known as Alexine. She was the daughter of a wealthy Dutch sugar trader with Liverpool connections, but was born in The Hague. Her father died when she was ten-years-old and the heiress became a very rich young woman. This gave her the necessary funds to be able to travel. Energetic and curious, she set off, chaperoned by her mother, Baroness Henriette van Capellen and her aunt, Adriana (Addy). She made her first visit to Egypt at the age of nineteen and from that time on Africa was to become her passion. She made three voyages up the Nile and got as far as Gondokoro.

Travelling up the Nile beyond Khartoum was extraordinarily dangerous because of the hostilities between Turkish-Egyptian authorities, Arab traders and the non-Arab native population. Even for an experienced Africa traveller it was difficult to prepare oneself for the dangers that could be encountered, as it was unclear from which side and to what extent hostilities might break out. The area was also fever-ridden. The great explorer, John Hanning Speke, gave a warning about the difficulties that might be encountered by the Dutch ladies and their team.

A subsequent trip was to a tributary of the White Nile, the Bahr el Ghazal (the Gazelle River). Money was no constraint for Alexine, and in addition to boats, she took five hundred porters, a train of donkeys and sixty five soldiers. They were accompanied by two German scientists, Baron Theodor von Heuglin (1824–1876) and Hermann Steudner (1832–1863). Their role was to map and to gather and assemble information on the geography, flora, fauna and people of the area. They reached Wau. It was a disastrous journey on which her mother, her mother's maid, her own maid, and Steudner all died. When the remnants of the party finally made it back to Khartoum, her aunt also died. She had, however, with the help of her German colleagues, made useful ethnographical and botanical collections.

In due course, Alexine went to Algiers, and planned thence to go southwards. Her caravan, consisting of over sixty camels and three horses, finally left Algiers in November 1867. Passing through Djelfa and Laghouat, she reached the walled towns of the M'zab at the beginning of 1868. She travelled in a large palanquin atop a camel. However, beset with arguments between the crew, delays and, eventually, harsh weather conditions, the journey became problematic and had to be aborted. She returned to Constantine via Touggourt and Biskra.

She then decided she wanted to be the first European woman to cross the Sahara. Her plan was to travel from Tripoli to Murzuk in Fezzan and thence to Lake Chad and Bornu, Wadai, Darfur and Kordofan to the Upper Nile. She settled in Tripoli in October 1868. While there she met several German explorers, including Gerhard Rohlfs. Ignoring Rohlfs' warnings about the physical and tribal dangers inherent in attempting to cross the Sahara, as well as the personal dangers due to her palpable wealth, Alexine organized a very large expedition. She succeeded in hiring two Dutch sailors in the hope that they might provide protection. Her caravan consisted of 70 camels, a large amount of luggage, including several large drums of water, and a train of over 50 people. It departed from Tripoli on 28 January 1869. She reached Murzuk in five weeks, and there she met up with the German explorer Gustav Nachtigal. She intended to cross the desert with him. Nachtigal respected Alexine for her experience, 'lofty aspirations and resolute will', but expressed doubts about her 'seemingly cold features and her distinguished aloof demeanour'. He also worried about 'her ample means and numerous motley entourage'. However, as Nachtigal wanted to go to the Tibesti Mountains first, she set out for the south on her own. Before leaving Murzuq her company had swelled to consist of two Dutch sailors, three Sudanese servants, three female servants, ten male servants (with wives and children), fifteen camel drivers, one Tunisian, two local inhabitants from Murzuq, three male slaves and one female who had run away from their proprietors and found asylum with her, and some more ex-slaves. Almost all of them were, according to Rohlfs, hanging around because of their interest in her as a potential benefactress, and not because they would be useful for this perilous journey. Her over-blown caravan advanced slowly, but on 1 August 1869, to the west of Murzuk, en route to Ghat, she was killed by the Tuareg. They cut off her hand, shot her and left her to bleed to death in the burning sun. Her two Dutch sailors were also killed. She was only 33.

Opinions vary as to Alexine's significance and character. To some she was seen as being dominating, stubborn, thoughtless, non-studious, spoiled and naive. To others she was seen as a brave, intrepid woman, who visited places that were largely unknown to Europeans, and made valuable collections. She was, perhaps, all of those things.

## HENRI DUVEYRIER

Henri Duveyrier (1840–1892) (**Figure 2.22**) was born of mixed English and French parentage in Paris. He was the eldest son of Ellen Claire Denie, an Englishwoman from Bath, and Charles Duveyrier, a well-known playwright. He had a very brief and very precocious career as a Saharan explorer, and is most notable for his respect for, and ethnographic descriptions of, the Tuareg. In 1857, having been educated in France, England, Italy, Switzerland and Germany, he went to Algeria. Here Henri met Oscar MacCarthy (1815–1894), a well-travelled and knowledgeable colonist of Irish descent. Eager to explore his new environment, Henri persuaded MacCarthy to organize a desert excursion and in March 1857 the two set off on a five-week long round trip to Laghouat. This five-week journey left a powerful mark on the young traveller and it was on this short tour that he encountered the Tuareg, 'the people of the veil'.

He then visited Heinrich Barth in London to seek advice and information so that he could undertake a more ambitious and extensive journey. The two men became friends and Barth generously afforded Duveyrier access to all his notes and materials. In 1859, aged only 19, he once more set foot in Algeria, at Constantine, and became the first European to visit the oasis of El Golea. He also visited Gadames, Gat and Murzuk in the deep south of Fezzan (Libya). In Constantine he donned Arab

**Figure 2.22** Henri Duveyrier. *Source: Alexandre Quinet, Public domain, via Wikimedia Commons.*

costume for his expedition. The early days of his voyage were extremely difficult, for he was attacked by bandits and had a potentially lethal fracas with two treacherous guides. At the end of his expedition, news spread of his exploits and Duveyrier became the hero of the French press. In January 1862, at the age of just twenty-one, he was awarded the Legion d'Honneur. As he recovered from his journey, he began work on the 70 notebooks he had kept during his voyage. These formed the basis of his acclaimed 1864 publication, *Exploration du Sahara: les Touareg du Nord.* This provided the first detailed description of the Tuareg. The book was much acclaimed and received the coveted gold medal of the Société de Géographie de Paris.

In it, Duveryier praised the virtues of the Tuareg. As Douglas Porch (1985, 73) wrote:

> Of all the European travellers, Henri Duveryier was the most impressed by the Tuareg. His description of the virtues of this noble race read like a cross between Sir Walter Scott and the Boy Scout Handbook: brave, hospitable, faithful, patient, tolerant of the defects of others, industrious, charitable, magnanimous, haters of oppression, and so on. The list is virtually inexhaustible. It was almost too good to be true. It *was* too good to be true.

Here are some quotes (translated) from his *Exploration du Sahara* (1864, 382 and 386):

> In general, the Tuareg are tall, some even seem real giants. All are lean, lean, sinewy, and their muscles seem steel springs . . . The Caucasian type is that of their faces: face oval and elongated in some, in others round: wide forehead, dark eyes, small nose, high cheekbones, average mouth, thin lips, white teeth and beautiful, where such have not been decayed by the use of natron, and rare black beard, black hair and smooth. Some have blue eyes, but this nuance is found infrequently. Eyes, in all people who are over forty years, seem veiled and obscure. This effect is due to the intensity of the light and to the action of solar reverberation. Many are blind or become blind before old age. The trunk, both in men than in women, is largely developed. The upper and lower limbs, elongated, muscular, ending in small and shapely hands and feet . . . Men are generally strong, robust, indefatigable, though their average power is much lower than the European . . . Women, too are big, have an arrogant carriage and are usually beautiful . . . Their appearance, however, is much closer to the European women than Arab women.

However, the book is not only about the Tuareg. The first 245 pages give a comprehensive account of the physical geography of the area, its hydrography, geology, meteorology, astronomy, vegetation, and animals (including lake crocodiles). The detail given, and the observations made are spectacular in quantity, quality and breadth, which is remarkable for one so young. There are, for example, descriptions of the different types of dunes, including linear seifs and star dunes, a discussion of the area they cover, and the direction of the winds that shape them. He also argued that the sand was the product of the intense disintegration of the desert surface brought about by temperature changes and other processes.

Duveyrier also wrote about camels and their value to the Tuareg (p.218–9):

> The life of the Tuareg, even more than that of other Saharans, is intimately linked to that of the camel; for this noble animal is not only its war mount, the locomotive of its caravan trains, *the express* which makes space disappear, this great enemy of the inhabitant of the desert, but also it is the provider of its principal needs. Its milk is almost the sole food of the family in the pasture season; Its meat is the *nec plus ultra* of the hospitality offered to the host of distinction; Its leather, one of the best in existence, provides the fabric of the tent, the raw material for saddles, shoes and most household utensils; Its hair provides the textile material for the lashing ropes of convoys; Its droppings, collected, are used here as fertilizing fertilizer for the palm trees; there, in large spaces without any vegetation, fuel with which to cook food; Finally, its trace, questioned in all the steps, provides the traveller with valuable information which is always taken into account, either because it announces the peaceful neighbourhood of a herd in grazing land, or because it signals the passage of individuals, isolated or in caravans, loaded or not, friends or enemies; for the width of the foot, the length of the nails, the nature of the droppings, reveal to the experienced man all he needs to know about the dispositions of those who follow or cross the same route.

Duveyrier made two more travels to North Africa in 1885 and 1886, and he gave advice to many Saharan explorers including Oskar Lenz and Renard Foureau, but he shot himself in 1892 with his own revolver in the woods near his home in Sèvres, having suffered from depression that may have been linked with French colonial policy and aggressive European imperial expansion. He was, as Michael Heffernan (1989, 349) suggested, 'a fearless explorer, an excellent natural scientist, an erudite anthropologist and a brilliant geographer'.

On the other hand, John Wright (2006, 310) regarded him less favourably. He believed that his writings on the Tuareg and the Sanussi were 'biased and dangerously misleading'.

## OSKAR LENZ AND CAMILLE DOULS

In the later decades of the nineteenth century, Timbuktu, located just north of the Niger River in what is now Mail, still held an attraction for European travelers. Two individuals who sought to make it to this fabled destination, coming from the north on camels, were Oskar Lenz and Camille Douls. One made it, but one did not.

Lenz (1848–1925) was a German-Austrian earth scientist who was born in Leipzig. In 1870 he earned his doctorate in mineralogy and geology at that city's university. In 1872 he joined the Imperial Geological Reichsanstalt in Vienna, and obtained Austrian citizenship. In 1879–80 he led his first trans-Sahara expedition, travelling from Morocco to Senegal. The main raison d'etre of the expedition was to perform geological studies including the possibility of locating iron ore deposits. With a Spanish Arabic-speaking companion, Cristobal Benítez (1856–1924), in 1880 he became only the fourth European to visit Timbuktu. The others had been Laing in 1826, Caillié in 1828 and Barth in 1853. Later his career he directed the Austro-Hungarian Congo Expedition, and crossed Africa from the Congo in the west to the Indian Ocean in the east. Following the completion of his work in Africa, he became a professor at the University of Prague.

With advice from Henri Duveyrier, and sponsored by the African Society of Germany, Lenz, disguised as a Turkish doctor, determined to explore Morocco and then to make his way southwards to Timbuktu via the small oasis town on Tindouf (**Figure 2.23**), the sand dunes of Iguidi and the salt mines of Taoudeni. He appears to have travelled in some style, carrying an expensive mattress and white sheets, and insisting on shaving each day. It was with a salt caravan from Taoudeni that he reached Timbuktu on 1 July, 1880. Here he was received with some civility, and after staying for twenty days, he took a route via Sokoto to Senegal, but was robbed and held to ransom, so that he arrived in the Atlantic port city of St Louis tattered, penniless and destitute. On his return he wrote a lengthy and meticulous account of his journey, entitled *Timbouctou*. Two years later it was translated into French, but there never was an English edition.

He provided important information on the dunes of Iguidi, and also heard 'singing sands' (p.59):

**Figure 2.23** Tindouf in 1880. *Source: Lenz 1886.*

> The long dunes of the Iguidi, which form entire chains with jagged crests at sharp angles, have, like all dunes, a surface slightly inclined towards the windward side, and another more rapid, and even sometimes very steep, of the opposite side. They also consist of a quartz sand, fluid, very pure, orange in colour, and made hot by the rays of the sun. When these hills are crossed by a caravan, there is a movement of small fluid and sonorous grains of sand; this movement, limited at first to a very small extent, soon occupies an increasingly large space and extends like an avalanche over the entire slope of the hill. The displacement of these grains results in making them strike against each other, which still produces a sound, although extremely weak; from the immense mass of grains of sand set in motion and the union of isolated sounds, however small they may be, then comes a noise which, in the Iguidi as in the peninsula of Sinai, can acquire a wholly quite extraordinary.

For the most part he loved the desert and its dunes (p.89):

> In the desert the atmosphere is of extraordinary purity and health; no diseases are known there, with the exception of eye ailments, which must be attributed to the uncleanliness of the inhabitants. I recommend hot sand baths in the dunes as a particularly beneficial cure for certain pains: it is a real pleasure to roll in the fluid and pure quartz sand, where there is not a speck of dust.

> The desert is beautiful, very beautiful, despite the heat and the dunes. There is something powerful and august about immense solitude which makes it analogous to the Infinite Ocean. A sunrise or moonlight in the Sahara has a charm that cannot be described; it is a spectacle of grandiose beauty, which produces unforgettable impressions. One who is capable of appreciating the great and the beautiful of nature, and who is endowed with a character fortunate enough not to be torn, by fear of possible danger, from the contemplation of all these wonders, the one he will certainly enjoy remembering the time spent in the Sahara, and will thank the happy destiny which will have allowed him to enjoy its beauties, with a healthy body and mind.

However, he disliked Araouan, a village in what is now Mali (p. 100):

> Each year, several thousand camels pass through Araouan; but a very large part come from the Taoudeni salt works; the latter do not pay, as far as I know, any rights. The water point of Araouan is therefore extraordinarily lively, as well as the pastures, located at a great distance from the city. The supply, maintenance and supervision of the camels which come to recover here after long journeys in the desert are the main occupations of the Rhatani, liberated Negroes. The presence of so many camels is also the cause of the most unpleasant afflictions in the area: flies. One cannot get an exact idea of the mass and the inconvenience of these swarms of insects, which one can only escape a little by standing all day in the darkest corners of the rooms. Add to that faulty food, lukewarm water, and the general unhealthy situation of the whole city, terrible heat, sand hurricanes as violent as they are suffocating, the absolute lack of any species of vegetation: these conditions come together make Araouan one of the hells of the earth.

He was beset by sand storms, but thought that their role in overwhelming camel caravans had been exaggerated:

> It is hardly necessary to say that the stories about the *samum*, this wind of death, which, it has been said, engulfs entire caravans, cannot be true. Such a hurricane may well cover animals and people with a thin layer of sand, but nothing more. It does not even seem possible to me that one could perish suffocated in a hurricane of this kind, because the true phenomenon lasts only a short time: each one protects his mouth, his nose, his ears and his

> eyes under a veil, by which certainly still penetrates a little sand, but which can be easily pushed aside afterwards. These hurricanes which cover and destroy hundreds of camels are part of the many fables written about the desert. It must certainly have happened that entire caravans were annihilated; but their disappearance was the result of lack of water.

From Timbuktu, Lenz, having tired of riding a camel, bought an ass and sometimes his companions even rode on oxen. Passing through lion-infested country, and meeting the Fulani tribes, Lenz eventually made it to St Louis on the Atlantic coast of Senegal.

Camille Douls (1864–1889) (**Figure 2.24**) entered what we now call Western Sahara at a time when the region was generally closed to foreigners, to live among the nomadic Ouled Delim, and to bring back valuable information on their way of life and environment. His first trip was in 1887. He meticulously prepared for this by staying in Morocco to learn Arabic, the customs of Muslims, and parts of the Koran. He was also circumcised. At the end of 1886, he left for the Canary Islands, opposite the Western Sahara (then called Rio de Oro). In January 1887 he was landed on the coast near Cape Garnet by Spanish fishermen, but was then quickly captured by the Ouled Delim nomads, who beat and stripped him. For days Douls, fettered hand and foot and lying on a camel, was carried along with the nomads, and subjected to searching inquiries by one sheik and another that they met, none of whom were fully convinced that he was telling the truth when he claimed to be a shipwrecked Algerian sailor. Eventually, having managed to pass himself off as a Muslim, he was accepted. He was thus able to spend five months with the Ouled Delim, whom he observed and questioned. He also went to Tindouf, which had previously been visited by Lenz. At last he became betrothed to a twelve-year-old daughter of a sheik,

**Figure 2.24** The striking and tragic Camille Douls, aged 25. *Source: Departmental archives, Rodez.*

and it was on the pretext of returning to Algeria to get a suitable marriage portion that he was permitted to leave the tribe and push north into Morocco by camel, going via Tafaya, the Atlas Mountains and Marrakech. There he was recognized as a Christian and imprisoned, but eventually released. He then returned to France, where he published his observations. His second and last trip was intended to be from Morocco to Timbuktu. He never reached his destination. Leaving Tangier, he crossed Morocco disguised as a pilgrim and passed through the Tafilalet to join the road to Timbuktu. He was assassinated by his Tuareg guides near Akabli, in the Adrar province of present-day south-central Algeria.

## FERNAND FOUREAU

The British sometimes suggested in the late nineteenth century that they had given the French the Sahara so that the Gallic Cockerel had a large expanse of sand to scratch in. In reality, however, the French did a great deal to appropriate this piece of terrain. It spawned some great soldiers and explorers who explored its arid wastes. Among these was Fernand Foreau (1850–1914) (**Figure 2.25**). He was born in the Limousin region of France. Later he was fortunate to study under Henri Duveyrier, the Saharan explorer, who helped develop Foureau's own interest in the subject. He undertook many expeditions, culminating from 1898 to 1900 with a long journey from Ouargla to Lake Chad with Army officer Amédée-François Lamy. This was known as the Foureau–Lamy Mission, and formed part of the French strategy to conquer Chad and unify all French dominions in West Africa. He made maps, established trans-Saharan routes and performed numerous geographical, geological and meteorological surveys. In 1899 he received the Patron's Gold Medal of the RGS.

He took his first trip to Algeria in 1876, accompanied by Lt Louis Say. The following year, he created the Oued Rhir Society, which promoted the digging of wells and the planting of palm groves between Biskra and Touggourt. Through this, he became familiar with Saharan customs and learned Arabic and some dialects of the Berbers. After the extraordinary massacre by the Tuareg of the Flatters mission of 1881, Foureau persevered and proposed to French authorities an exploration of the great desert. Between 1882 and 1897 he made nine further journeys into the Algerian Sahara. Using camels, he explored the Tademait, Tidikelt, the great Western and Eastern Ergs (sandseas), and the Tassili-n-Ajjer. In all he covered around 21,000 km.

**Figure 2.25** Fernand Foureau during the Foureau–Lamy expedition to Chad in 1900. *Source: picture from the French weekly newspaper* L'Illustration, *9 September 1900.*

By 1895, he was convinced that only a large-scale military gesture could ensure a permanent French presence in these regions, and that if the Tuareg could not be persuaded to open the door to the Sahara, then it must be forced. Upon his return to France in 1898, he was able to convince his sponsors of the need to undertake such a venture and received financial support from the Société de Géographie de Paris. The Foureau-Lamy mission (also called *Mission Saharienne*) travelled southwards from Algiers, and arrived in Agadez in July of 1899, and in Zinder in November of that year. It eventually went on to Chad.

Douglas Porch (1985) asserts that Foreau was made titular chief of the expedition to camouflage a mission of colonial conquest, for Foreau was a civilian, whereas his colleague Major François Lamy was a hard-headed and resilient military man who had seen long service in North Africa and the Congo.

Porch described Foureau's appearance (p.148–9):

> Foreau exuded the asceticism of the desert – a long head perched on a thin neck, and a tall, spare frame, made him seem as if he had been constructed for heat loss. His brown, wrinkled face had the appearance of a much-used chamois. Only a few crinkles of hair twisted up from his near-vacant scalp. There was nothing particularly decorative or handsome about Fernand Foreau. Like a camel, he appeared to have been designed for the sole purpose of survival in the dry regions.

Porch made a gloomy prognosis of the expedition's likely success, for most of its participants had meagre experience of dealing with camels, and the number of camels involved was huge (p.165):

> As the Foureau-Lamy expedition made its way south from Biskra to Touggourt and Ourgla, it grew like a snowball by acquiring tirailleurs and Chaamba cameleers until, at the jumping off point, it numbered 381 men. To cross a hostile desert with 381 inexperienced men and 1004 camels was like trying to maneuver a battleship in a three- acre lake with a crew composed of Sea Scouts.

Furthermore, the quality of the camels was mixed, and the saddles and harnesses were poor. They also had to cover no less than 3200 km between Ouargla and Lake Chad. Most of the members of the mission were African troops (tirailleurs) of slave extraction.

The camels were partly used as riding animals, but mainly as beasts of burden for the equipment, which consisted of 5000–6000 pieces of baggage. Among the items were tents, furniture, food, ammunition, guns and cannons, medicines, scientific instruments, barter goods, and so forth. The provisions alone were intended to last for more than a year. The burdens were too great and the conditions too savage. As the expedition moved south there was terrible attrition of camels. By the time they reached Agadez in Niger, only two of the original camels remained. Their loss made it necessary to abandon or burn nearly all their equipment. This included the European food supplies, scientific instruments, as well as household goods, such as tents and beds, which had provided the expedition with European comforts. Since there were no more riding animals, even the officers had to travel on foot.

As his obituary in the *Geographical Journal* (Anon 1914) remarked, 'By this great journey the way was finally opened for the extension of French supremacy across the Central Sahara to the Sudan, and the junction of the spheres of that nation in North, West and Equatorial Africa.' It needs to be added, however, that Foureau also accumulated a vast amount of material on the great desert he had so often traversed.

## JAMES HARRISON

James Jonathan Harrison (1857–1923) was born in Yorkshire and attended Harrow and Christ Church, Oxford. He then joined the Princess of Wales's Own Yorkshire Hussars in 1884 and remained with that élitist yeomanry regiment, experiencing no war service, until September 1904, when he retired with the honorary title of lieutenant-colonel (Green 2004). He is perhaps best known for bringing a group of six pygmies (1.2–1.4 m tall) back from the

Congo rainforest in 1905. These wretched Africans, armed with spears and poisoned arrows, went with Harrison to a garden party at Buckingham Palace and to Parliament within days of arriving in London. They were forced to tour the country as a commercial spectacle and were seen by a million people at such venues as the London Hippodrome and Olympia. They were returned to the Congo in 1908.

Before that, in late 1899 and early 1900, he undertook a camel journey from Zeila (Saylac), on the Somali coast, across to Lake Rudolf (now called Lake Turkana) on the borders of Kenya and Abyssinia (Ethiopia). This was a remarkable journey across the entire width of Abyssinia from north east to south west, and was written up in the *Geographical Journal* for 1901. Travelling with a group of Europeans, including a surveyor and a taxidermist, they induced Mohamed Hassan, a Somali 'boy', to go as their headman. They also employed a splendid group of 61 local 'boys'. They had 80 camels. They travelled by Gildessa (Jaldessa, Jeldesa) in eastern Abyssinia, and then reached the Hawash (Awash) river, where they slaughtered many elephant. On the Abyssinian Christmas Day they met the Emperor in Addis Ababa and provided him with a splendid lunch that included claret and champagne.

**Figure 2.26** Lake Stefanie in a dry state in 1970. Note the stranded spit. This is probably how Harrison saw it seventy years earlier. *Source: Author.*

They eventually reached Lake Zwai in the Rift Valley, where their camels ate some poisonous vegetation called *gumbot*. Some of them went mad and 30 of them died. They continued to Lake Abaya and on to Lake Stefanie (Chew Bahir) (**Figure 2.26**). It was a time of great drought, and this lake was dry (p. 271):

> We now saw ahead, as far as the eye could reach, nothing but a vast extent of ground strewn with shells and heaps of fish-bones. 'Stephanie-dry!' we all said; and a thunderbolt in our midst could not have caused more consternation.

The water they obtained from wells was dire, and over 40 of their boys lay in agony as a result of drinking it. They then visited Lake Rudolf and the Omo River, which was not flowing. They were not the first visitors to Stefanie and Rudolf. These had been visited and named by Count Teleki and Ludwig von Höhnel in 1888 (von Höhnel, 1938; Grove and Goudie 1971), and had been visited by two British parties travelling in from Somaliland (Smith 1898; Cavendish 1898).

From there they entered Northern Kenya and noted some of the local people near Mount Nyiru (Harrison 1901, 274):

> On reaching Mount Nyirobuk, we encountered the Boma and a tribe called Lokuba. Clothing was quite dispensed with; slitting the ear, they insert a large round piece of wood, or some swells prefer four brass cartridge-cases. Hairdressing is practised in many styles, some having it down to their waist, but small curls smeared and plastered down with a red brick-dust looked best. They wear iron armlets and beads, live by hunting, and use bows and poisoned arrows, with long spears. The next tribe, called Semirides, own large flocks of sheep and cattle. They are a well built and good-looking race.

By May they were at Lake Baringo in the Kenyan rift valley and then at Naivasha, where they reached the Uganda Railway. Here they said goodbye to their camels and took a train to Mombassa. They had travelled c.2300 km at an average of 11.5 per day. One camel had carried a 280-lb load of rice through from Zeila without a break. The team had made many observations of birds and collected their skins. They also recorded that on the journey they had hunted extensively and had shot 381 head of game (Grant and Harrison 1901).

Captain Montagu Sinclair Wellby also used a caravan to reach Lake Rudolf in 1899 (Wellby 1900), before being killed the following year in the Boer War. Philip Maud and Archibald Butter made a similar journey by camel from Dibouti to Lakes Stephanie and Rudolf in 1902–3 (Maud 1904). It was an intelligence gathering trip, but they too managed to slaughter many lions, leopards and the like.

## RENE CHUDEAU AND CONRAD KILIAN

Some geologists played a major role in exploration in the Saharan and Libyan Deserts (Taquet 2007). These included René Chudeau (1864–1921) and Conrad Kilian (1898–1950) in the former, and Ardito Desio in the latter.

In 1904, Emile-Félix Gautier (a geographer and author of *Sahara*) asked a 40-year-old associate to accompany him. This was René Chudeau, a small and vigorous geologist. They met up at Timimoun in south-central Algeria in May 1905. The project was to establish the route of a telegraph line between Salah (southern Algeria) and Timbuktu (Niger). This led Chudeau to cross the Sahara from Zinder in Niger and Lake Chad. Eighteen months after his departure he returned to France via Bamako and Dakar (Senegal). He had covered a distance of 7500 km (Gautier and Chudeau 1908). He mapped the districts he crossed, made a geological survey and took numerous observations. Among his achievements was the discovery the first dinosaur bones in Niger at the bottom of the Tiguidi cliff, south of Agades.

Chudeau started again in 1908 for Mauritania and in 1909 travelled in the Timbuktu region. At the end of the same year, he crossed the Sahara again from north to south. He returned in1913 and 1914 to Timbuktu and visited the Taoudeni basin, discovering unknown ancient volcanoes. In all, on his trips Chudeau walked a total of 18,000 km through unexplored areas of the Sahara. Chudeau demonstrated the presence of Cretaceous rocks in the south of the desert. He was also interested in the Quaternary deposits and in geomorphological phenomena, such as the formation of dunes, the patina of the rocks, and the processes of wind erosion, and the possible capture of the Niger River by the ancient Taffassasset. Indeed, his work on dunes, and the classification of all the dune types he encountered, was a remarkable contribution. However, back in Paris, Chudeau was obliged to accept a very modest job as a consulting engineer and died miserably and alone.

François Théodore Conrad Kilian was born in August 1898 in the Ardeche region of France. His father was Wilfrid Kilian, an alpine geologist. In October

1921 Kilian was engaged by a businessman living in Constantine (Algeria) who wanted to find the legendary locality of the emerald deposits of the Garamantes, ancient inhabitants of the Fezzan area of southern Libya and neighbouring parts of the central Sahara. He was employed as a prospector and embarked on his survey in the company of an older warrant officer of the camel corps. After many incredible mishaps, Kilian, who split with his military companion, Lacroix, completed his geological exploration of the Tassili n'Ajjer and of the central Hoggar in southern Algeria. Fascinated by the Sahara, he returned again in 1926, when he crossed the desert on a white camel dressed like a Touareg. During this trip he made numerous discoveries; he established the existence on the Saharan platform of a 'Continental intercalaire' containing fossil reptiles and fishes. It was also Kilian who defined a 'Continental terminal' of Upper Cretaceous and Cenozoic age.

At the beginning of February 1928, Kilian reached Tamanrasset to prepare another expedition to cross the Ténéré from west to east. He was rewarded with the gold medal of the French Geographical Society and he was invited to London in May 1932 by Francis Rodd to deliver a lecture on his Saharan explorations to the members of the RGS. At the beginning of 1935, Kilian returned to the Sahara, charged by the French government to examine and help define the entire border area with Libya and Niger. He also developed an interest in the geology of Fezzan in Libya. He was, in short, one of the first people to recognise the hydrocarbon potential of North Africa. Kilian died in a mysterious and strange way in a hotel in Grenoble in 1950. He was found hanging and covered with blood. There is uncertainty as to whether this was a case of suicide or of murder. He was aged 52, the same age at which Henry Duveyrier killed himself.

## ANDRE VON DUMREICHER

From 1896 onwards, a Camel Corps section of the Coastguard Administration in Egypt was employed to oversee the deserts around the Suez Canal and to prevent the smuggling of salt (on which there was a state monopoly), tobacco and hashish. The Camel Corps was also charged with warding off the illicit landing of pilgrims on the Red Sea coast and to patrol Egypt's borders, including that with Tripoli. The Corps was headed by André von Dumreicher (1865–1953) who was a descendant of a German-Danish family that had formed part of Alexandria's European community since the late eighteenth century. They had grown cotton and owned cotton mills, and it is said that Napoleon stayed at their house during his Egyptian campaign.

Von Dumreicher, who was born in 1865, commanded about 500 cameleers whose task it was to secure a border of more than 4000 km in length. The Nile Valley had by then been brought under quite close administration, but the deserts on either side, comprising 96 per cent of Egyptian territory, had been in part untouched by the government until the Camel Corps (**Figure 2.27**) was charged with its surveillance.

His book, *Trackers & Smugglers in the Deserts of Egypt* (1931), which was simultaneously published in German as *Fahrten, Pfadfinder und Beduinen in den Wüsten Ägyptens*, is remarkable because of what it has to say about camels, how they could be used in tracking, and how they could be tracked. He also gained unparalleled knowledge of, and sympathy for, the Bedouin.

Dumreicher noted the importance of camel caravans for the date harvest at Siwa oasis (p.23–4):

> During the date harvest the transport of this fruit by caravan is a profitable undertaking for the nomad; it is, in fact, the great event, and for most Bedouins the only long caravan journey of the year. During the height of the season 2,000 bedouins, with 15,000 camels, gather near Siwa, and even the camels feed on dates … From Siwa the great caravans of dates go off in every direction. The tribes of Tripoli undertake the transport to the west, while the caravans of the Aulad Ali proceed to the Mediterranean coast and to all parts of the Rif.

**Figure 2.27** Men of von Dumreicher's Camel Corps. *Source: Dumreicher 1931.*

Dumreicher (p.72–3) wrote at some length about the relationship between male and female camels (*nagas*):

> In the desert the animals' sense of smell is very acute and they scent their mates at a great distance. Attracted by the smell of our stallions many young nagas came trotting up, but our young camels who hardly knew that mares existed, were perplexed and confused. The young mares knew little more of the world, but their curiosity conquered their shyness. They were healthy, beautiful animals and had, as all the creatures of the desert – bedouins, gazelles, and ostriches – beautiful and patient eyes....our camels soon began to be interested in the young mares. They cautiously rubbed nostrils and pursed their lips, and bending their necks backwards after the practice of the turkey, tried to look as attractive as possible.

However, the most interesting part of Dumreicher's book is when he discusses the tracks of camels, and how they indicate the individual beast, their age, their loads, and their rate of travel. This information is crucial when in pursuit of thieves and smugglers, or when searching for laggard or straying members of a flock. As he wrote (p.94):

> Every track of man and beast on the few caravan roads which exist interests the coastguards. All tracks in the open wastes where there is neither grazing nor water are highly suspicious. Instead of obtaining the news of the day by reading his paper, the coastguard officer lets his guide interpret the latest desert news by the signs written in the sand. It is an axiom in the desert that a man's conscience is reflected in his tracks. The desert is eternal, the camel is slow, the Bedouin, like the Sphinx, disdainful of time. Therefore the least indication of speed in the tracks raises suspicion. Smugglers only choose the caravan road in the hope that their tracks will soon be obliterated by the tracks of following caravans....the tracker is instinctively aware of the dissimilarity of the straight trail of the led or driven smuggler camels and those of a caravan meandering while grazing on the march from bush to bush.

Dumreicher also discussed camel dung and what could be learned from it (p.143):

> The camels we were following must have been fine, big and healthy animals which had been fed on clover, chopped straw, and even grain. The traces of smell of odoriferous plants like wild garlic, sage, thyme, or mint

> were still perceptible in the animals' dung. From all these clues Ahmed concluded that the tracks were those of camels belonging to the Belly, a small but wealthy sedentary tribe from Nag Hamadi, and that they had grazed on the way in a certain wadi where wild thyme abounded.

Even the number of camels could be calculated by the number of heaps of droppings that there were present at a water-hole, and the presence of baby camels was indicated, not surprisingly, by particularly small heaps.

Dumreicher was solicitous for the welfare of his camels, and tried to protect them from ticks, mosquitoes, camel flies and the mange. He also designed a new type of camel saddle so that back sores were eliminated. Dumreicher's apprentices, including Leo 'Bimbashi' Royle, continued his work in the Coastguard, but the days of using camel patrols to chase smugglers and watch the frontiers were numbered. The motor car and the aeroplane were to largely replace them. In addition, as Russell McGuirk (2007, 67–8) has revealed, in late 1914, in Sinai, a unit of the coastguard may have deserted, causing the death of members of a patrol of the recently arrived Bikaner Camel Corps.

He was head of the Corps until 1910. In the First World War, being seen as an enemy alien, he was given the choice of being interned in Malta or in England. Dumreicher had an English wife who was niece of a major British administrator in Egypt, Sir Eldon Gorst, so he chose England and was sent to Alexander Palace. He died in 1953.

## W.J. HARDING KING

William Joseph Harding King (1869–1933) was educated at Jesus College, Cambridge. The results of his first journey of exploration in the Western Sahara in 1900 were given in *A Search for the Masked Tawareks* (Harding King 1903). Harding King departed with a small caravan from Biskra in Algeria in order to visit the Tuareg of Hoggar. This was the first account of the Tuareg in English since Barth's work had appeared more than half a century before.

Following the death of his father, in 1903, Harding King appears to have inherited funds to permit his travels. Further exploration followed in the Sahara in 1908 and from 1909 until 1912 in the central parts of the Libyan Desert. In 1913 he joined Oric Bates, a great expert on the anthropology of eastern Libya, in the archaeological exploration of Marmarica in Libya. His most important book, *Mysteries of the Libyan Desert*, appeared in 1925. In this he gave an account of his penetration to the central Libyan Desert, notwithstanding Senussi opposition. This contains valuable geographical, ethnological and archaeological material. Harding King contributed many papers on his investigations to scientific periodicals, including the *Geographical Journal*, and was awarded the Gill Memorial Medal by the RGS in 1919. In summary, over a period of three decades, Harding King was one of the greatest explorers of both the Sahara and the Libyan Desert, contributed to debates about the mythical oasis of Zerzura, and also did some important scientific work on sand dunes and ripples.

In *Mysteries of the Libyan Desert,* Harding King describes a journey from Kharga in 1909. Arriving by train, he recruited a guide, Qway Hassan Qway, and obtained camels. He decided to travel westwards to Dakhla Oasis and the great dune belt of Abu Moharik. His description of Qway provides an illustration of what a guide looked like and needed (p.33):

> Qway's equipment was about as near perfection for the desert as it was possible to get it. His camel saddle was a *rabiat*. Over this was a red leather cushion on which he sat. On this he placed his *hurj*, or pair of saddle-bags of strong carpet-like stuff, one of which hung down on either side; above this lay a folded red blanket, and over this again he spread his *furwa*, or black sheepskin – an indispensable part of a camel-rider's equipment, which he not only places over his saddle, where it forms a soft and comfortable seat, but on which he sits when dismounted, lies on or covers himself with at night and throws over his shoulders on a cold day. Over the camel's withers in front of the saddle was a second small pad, also of red leather, on which to rest his legs as he crossed them in front of him as he rode. Hanging from his *rabiat* on either side was a sack of grain for his camel, the pockets of his *hurj* resting on top of the sacks.

In addition, Qway carried a Martini-Henry rifle, a red parasol, a goat-skin for carrying water, a small crock full of cheese, his camel's nose bag, a few eggs packed in straw, a camel hobble, and a skin of flour.

Some of the camels became severely emaciated on his journeys (**Figure 2.28**). Until the camels got used to each other they could spend time fighting. Harding King was not one to tolerate bad camel behaviour (p. 35):

> A savage camel is a dangerous beast and it is of no use playing with him. The right place to hit him is his neck. Hit him hard with something heavy and go on doing it and he becomes partially stunned and is then amenable to reason.

As to his own mount (p. 36):

> He was not a bad mount, but as he had not been ridden before I bought him, and guiding a camel by means of a single rein is always rather like trying to steer a boa-constrictor with a string, my stick at first had to be used pretty often.

Strict treatment of a camel was not inevitable. One of his guides played music to soothe the camels on a reed flute, and this proved to be effective, though the concert kept Harding King awake at night.

In the Kharga oasis, Harding King surveyed the crescentic barchan dunes that threatened the new railway, and calculated the speed at which they could migrate (*Geographical Journal*, 1918). In addition to this visit to Kharga and Dakhla, Harding King also went to Farafra (see Harding King, *Geographical Journal*, 1913). Here he discussed the remarkable wind erosion which created bizarre forms in the limestone bedrock (**Figure 2.29**). In 1910–11 he seems to have seen, but not visited, the Gilf Kebir.

Harding King did a great deal of work of practical value, including seeking sources of water and strategic information (Thomas and Hill 2009, 87):

> In addition to his mapping-related survey work Harding King undertook a wide variety of collecting and observational activities throughout his travels. He collected head measurements and ethnographic data of people

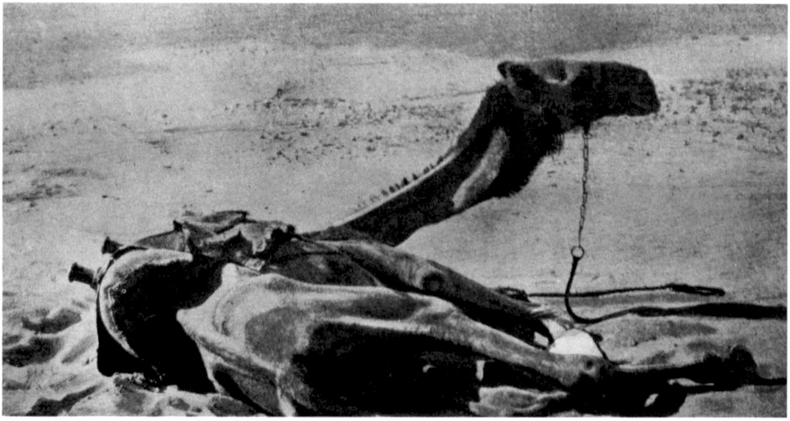

**Figure 2.28** One of Harding King's camels in an emaciated state. *Source: Harding King 1925, 181.*

**Figure 2.29** Wind-moulded limestone in the white desert – Farafra. A photograph, possibly of this very feature, is given in Harding King's paper (p. 456–7). *Source: Author.*

> 'who have not I think been done before', alongside oral folklore, songs and population data. In addition, he made botanical, zoological and geological collections; copied rock inscriptions; noted camel brandings; and made descriptions of settlements and dwellings and the encounters he had with people in them.

Towards the end of his career, Harding King engaged in debates with Francis Rodd, John Ball and others about the meaning and location of Zerzura (*Geographical Journal*, 1928, 1930). He may not have been an easy man, and his views on the local people he met are far from flattering. As Thomas and Hill (2009) have demonstrated, 'Harding King emerged as an author who was highly sensitive to criticism and would defend himself against an unfavourable review with dogmatic belligerence'.

## CHARLES FURLONG

In the early twentieth century, various explorers visited Tripoli and travelled by camel from there along the coast and into the interior. One of these was the French archaeologist, Vicomte Mehier de Mathuisieulx (de Mathuisieulx 1904). Another was Charles Wellington Furlong (1874–1967) **(Figure 2.30)**.

**Figure 2.30** One of Furlong's drawings of a camel caravan in Tripolitania.
*Source: Furlong 1909.*

He was an American explorer, soldier, writer, artist and photographer from Cambridge, Massachusetts. He graduated from Massachusetts Normal Art School in 1895. He was also a student at Cornell, Harvard and the Ecole des Beaux Arts in Paris. In 1904–5 he visited North Africa, and spent time in Tripoli and Tripolitania, possibly the first American to do so. As a result of his experiences there he wrote a book that appeared in 1909 and was reprinted in 1914. It was called, *The Gateway to the Sahara, Adventures and Observations in Tripoli.*

Furlong encountered camel caravans (Figure 2.30) (p.176–7):

> Caravans vary in size, from that of some lone nomadic trader or esparto picker, who trudges beside his few camels on his way to some local market, to the great trans-Saharan trade caravans with thousands of camels, not to mention donkeys, goats, sheep, and dogs. Such a caravan is rarely met with; it takes a year or more to outfit; thousands of dollars are invested by Arabs and Jewish merchants. Its numerical strength is increased by smaller caravans, whose sheiks, believing in the safety of numbers, often delay their own departure for months. Moving south from Tripoli, it must cover some fifteen hundred miles of arid Desert before it reaches one of the important marts of the Sudan.

> After numerous stops and leaving many animals and some men to the vultures, the caravan, if fortunate, reaches its destination. In its heavy loads are packed the heterogeneous goods generally taken, consisting of cotton and wool, cloth, waste silk, yarn, box rings, beads, amber, paper, sugar, drugs, and tea, of which British cotton goods form more than fifty per cent, of the value. Besides these it carries some native products. This cargo is bartered for the products of the Sudan: skins, ivory, ostrich feathers, guinea corn, and gold-dust. Every autumn caravans also arrive from the interior and return with dried dates; for, among the tribes of the Fezzan, Tripoli dates form the chief article of diet, and in the oases of the Desert, dates chopped with straw are used as fodder. A year, perhaps, after its arrival it begins the return voyage, with a cargo likely enough amounting to nearly a million dollars in value; and it is a gamble whether it ever reaches Tripoli.

After leaving Tripoli he developed a fondness for the camel (p.208):

> Often in the narrow streets and open suks of many North African towns I had met the great lumbering jemal, but it was not until I had eaten in his shadow, slept by him in fonduks, and travelled with him day by day along the caravan trails of the northern Sahara, that I began to understand and fully appreciate this incongruous model of ugly usefulness.

He also compared the different types of camel (p.211):

> While there are many different breeds of camels, the most distinctive of the Arabians are the heavy, slow-moving jemal and his cousin, the mehari, a tall, lightly built, swifter, and more elegant creature, used almost exclusively for riding, known as the riding or running camel.
>
> One has but to try the experiment of riding a baggager to realize that there is not only a distinction with a difference, but a distinction with a vengeance; and any Christian who willingly substitutes the rump of a jemal for terra firma deserves all he gets, for even an Arab will often prefer walking to the lumbar vertebrae of a jemal.

He remarked (p.214) that the camel with his stiff legs and supple neck was 'a combination of serpent and lamp-post'. He also admired the temperament of the camel (p.219):

> At times, perhaps from the heat, or without apparent cause, the camel is seized with a terrific frenzy; then look out — for he will attack driver, other camels, or any living thing. But, on the whole, this ancient burden-bearer of man is a dignified, long-suffering, lugubrious anomaly.
>
> He is not aggressive; his indifference toward man seems almost contemptuous. He is imperturbable and patient beyond precedent, and on the march will continue to stagger on until his last ounce of strength has been exhausted.

After his experiences in the dry, hot wastes of the Sahara, this gifted artist and writer travelled to the wet, cold area at the southern tip of the Americas – Tierra del Fuego. The environment could not have been more different.

## JEAN TILHO

Jean Tilho (1875–1956) was a French army officer, explorer and administrator who vastly increased knowledge of the southern Sahara, and also of Lake Chad. He left Saint-Cyr Military Academy in 1895 to join the Colonial Infantry and immediately participated in border demarcation missions in Niger. In 1903–1904, he was a member of Henry Moll's Mission, which worked on the French Niger and British Nigeria borders. In 1906–1907, he headed a politico-scientific mission in the same sector. In 1908, he visited the Chadian Bahr El Ghazel and Lake Fitri. That great soldier, Victor Emmanuel Largeau, put him in charge of Kanem in 1912. He then participated in military operations in Borkou and Ennedi and studied climatology, natural sciences and ethnography in the border areas of Chad, the Anglo-Egyptian Sudan and Libya. In 1914 Largeau installed him as commander of Borkou-Ennedi. He maintained this position until 1917 and fought against the Senussi.

His scientific work was considerable. He demonstrated and mapped the exact location of Lake Chad, established that there was no direct link between the Lake and the Nile, that meteorological fluctuations had caused periodical variations in the extent and levels of the Lake, and that the lake could be endangered due a capture of the Logone-Chari system by the Benue River. His map of Chad provides, together with his detailed descriptions of its different components, such as channels, marshes and lagoons, a benchmark against which its subsequent fluctuations in response to droughts and other factors can be gauged (Tilho 1910–1911 and 1910) (**Figure 2.31**).

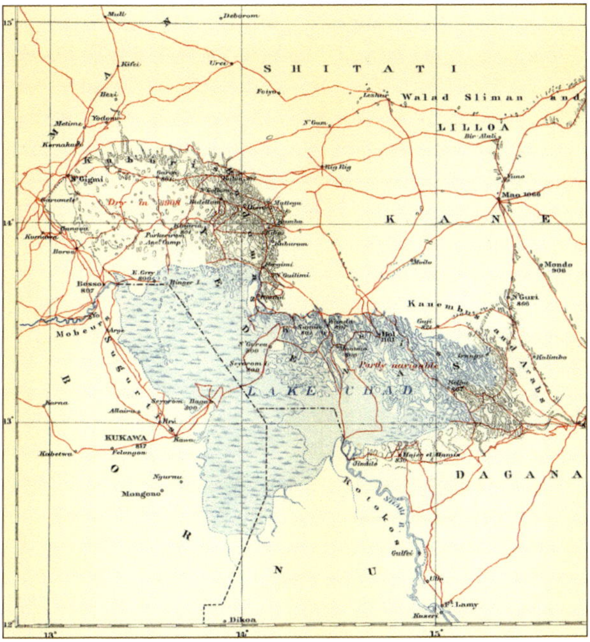


**Figure 2.31** A portion of Tilho's map of the Chad basin, showing its state in 1907.
*Source: Tilho 1910.*

In addition, between 1912 and 1917 he carried out over two thousand kilometers of highly detailed mapping of an area stretching from Tibesti to Darfur and the Nile (Tilho 1920), much of it using camel transport, and visiting such places as Ennedi, Bourkou, and Erdi. The motives were a mix of science and the military subjugation of territory and tribes for the glory of France.

On p.84 he discussed his caravan en route to subdue the Senussi of Borkou:

> Our expedition consisted of 400 black soldiers, with two mountain-guns; about 200 Arab and Toubou volunteers, forming a 'goum' or party of scouts, accompanied the column.
>
> We carried with us provisions for forty days, and the total number of our camels was about 2000.

From Faya, they went on to the western borders of the Libyan Desert, and he was one of the first Europeans to describe Ounianga (p.92):

> I reached the valley of Ounianga on October 9 in the morning, and was not a little astonished at failing to see the palm plantation till the moment of entering it; for, unlike those of Borkou, which can be seen from a distance, the oasis of Ounianga is hidden in a rocky excavation some 30 yards in depth and 4 or 5 miles long by 1 or 2 wide. The landscape thus formed is incomparably picturesque: a great sheet of calm water with blue shadows, edged with rosy-tinted beaches of sand, and fringed with green palm-trees stretched within a circle of bare wind-carved sandstone whose sombre hues cast here and there, under the blazing sun, warm shadows glowing with red or gold.

From here they went to Tekro, located in the north of Chad in close proximity to the borders of Libya and Sudan, and south of the Sarra triangle (p.94):

> The well of Tekro is particularly important, because it is situated at the extremity of the great caravan route joining the Mediterranean to the Soudan by the oases of Koufra and the well of Sarra. The water is abundant and fairly fresh, but the vegetation is reduced to a hundred clumps of siwak and a few tufts of grass of no value for the feeding of camels.

They also went to the upland mass of Erdi (p.97):

> In two marches we succeeded in reaching the foot of the cliffs of Erdi-Dji, 750 feet high and about 2000 feet above the sea. We found there good grazing for the camels, and from that day onward we had abundant fodder at each successive stage, so that I was delivered from the dread of seeing my indispensable beasts of burden waste away from inanition. The water was no less abundant, and was found in natural cisterns hollowed out by waterfalls in the beds of dried-up torrents that came down from the plateau. Some of these cisterns contained nothing but sand; but it was enough to bore a hole 1 or 2 feet deep in the sand to obtain a sufficient store.

Then it was on to the great sandstone massif of Ennedi, with its lakes and streams. It was difficult country through which to travel (p.163–164):

> For a week our route lay through a maze of sandstone rocks where no track existed, and through which our guides zigzagged from crest to crest with remarkable sureness. Sometimes we made a long detour to cross a wadi near its source; sometimes we marched straight for the

> obstacle, dropping down steep ledges that inspired little confidence in our animals, or crossing difficult ridges that the camels could only climb after being unloaded. Everywhere were narrow gorges and jagged crests, with here and there a few leagues of easy going in the neighbourhood of the temporary pools that usually marked the convergence of certain important ravines.

In September, 1916, he visited Tibesti, and climbed to see its great volcanic crater.

In 1917 he was involved in further military operations. He went from Borkou to Wadai, visiting Abéché. This city had lost much of its former grandeur, but was lively enough (p.246):

> Men, women, and merry black small-fry bustled noisily to and fro, inextricably mixed up with asses, camels, dogs, and horses. Numerous Tripolitan merchants, white-faced, wearing red fezzes and long flowing embroidered robes, stalked gravely back and forth, making it evident by their decorous elegance and the satisfaction visible on their faces that, in spite of the suppression of the slave-traffic, business remained active and prosperous

Finally he reached El Fasher in the Anglo-Egyptian Sudan. He regarded this town with great appreciation, noting is its well-kept and green appearance, wide streets, and houses in good repair and surrounded with trees. He then went by camel to the rail head at El Obeid, and, enjoying the luxury of the train, returned to France via Khartoum and Cairo. From Paris he went to the less luxurious Western Front.

## HANNS VISCHER

Hanns Vischer (1876–1945) (**Figure 2.32**) is a name that is now largely forgotten in the annals of exploration, yet in 1909, in the *Illustrated London News,* he was considered to be an explorer on a par with Shackleton.

**Figure 2.32** Hanns Vischer in c.1909. *Source:* Le Mois Géographique, *2 May 1909.*

In 1906, Vischer embarked on a remarkable 2400-km journey with 1500 camels through the Sahara, from Tripoli to Kukawa in Bornu, northern Nigeria. In 1910 he published *Across the Sahara*, an account of the journey illustrated with his own sketches. That he is now once again remembered is largely because John Hare retraced his steps by camel across the Sahara (albeit in the opposite direction) and wrote up his experiences in his dramatic *Shadows across the Sahara* (2003).

Vischer was born in Basle, being descended on both his father's and mother's sides from Huguenot refugees who had fled France in the seventeenth century. He had his early education in Basel and at Niesky in Germany. Later, he travelled to England and attended St Lawrence's College in Ramsgate before earning his degree from Emmanuel College, Cambridge. Towards his final year in college, he had developed interest in African ethnology and had attended a Hausa language course in Tripoli. After leaving Emmanuel College, he spent a year at Ridley Hall where he became familiar with the Church Missionary Society (CMS). In 1901, he travelled to Nigeria with some of their missionaries and visited Hausaland. His stay in Lokoja was limited as he suffered several bouts of fever, and he went back to his native home to recuperate. He returned in 1903, as a naturalised British citizen, resigning from missionary work and joining the colonial service as an Assistant Resident. His first posting was in Bornu, in 1905. A keen linguist, he became proficient not only in Hausa, but also in Arabic and a number of other local languages. In Bornu he heard about the caravan trade and the slave routes between Tripoli and Bornu. This stirred his desire for his journey. He left Tripoli in July, 1906, as an officer of the Government of Northern Nigeria to join his post in Bornu.

Vischer assembled a caravan in palm groves outside Tripoli. The hired camels came from the Megarha, a Beduin tribe living on the northern border of Fezzan, and were used to carry his loads as far as Murzuk. There he bought some and hired camels from the Tubbu and used them as far as Bilma. From Bilma to Bornu he was allowed to march with a French expedition, whose commandant, Martin Henri Jacques Gadel, very kindly supplied him with transport. When he started off he had a rum assortment of participants. As John Hare (2003, p.6) remarked:

> Vischer mingled in his caravan the most combustible materials he could possibly have collected. There were the Mecca pilgrims, who insisted on being escorted back to their homes in northern Nigeria; the freed slaves, who were to be his armed escort but who more often themselves

> needed escorting; turbulent and treacherous Tubbus from Niger, and haughty Arab camel drivers, who regarded black men as dirt and Christians as reliable fuel for hell fire.'

He also had to take with him the men's women and children, an accessory which does not help a traveller across the desert when each meal and every drink of water must be calculated. Together, they travelled along the old slave caravan route across the dreaded Hamada el Homra, to Murzuk in the Fezzan, on to Bilma, and thence to Kukawa near Lake Chad. It was a route which in the nineteenth century claimed the lives of many European explorers.

Things started none too well. The noise of drums, howling women, and cursing Arabs, caused his camels to bolt. Vischer was particularly struck by the noise that could be made by some of the women (p.51):

> I have seen the ladies of French and Italian fish-markets argue, have watched fights between their dark cousins on the Niger and the Benue; I have heard the language of excited Jewesses at Tripoli, but none equalled these back beauties when they were roused.

On one occasion (p.73) one of his 1500 camels was startled in the night:

> . . . one of our camels suddenly went mad without any apparent reason. From the midst of the sleepy caravan, the animal started up with a snort and a howl, gave one frightened look backwards and charged at a mad gallop into the disconcerted column, flung his loads right and left like stones from a catapult, and disappeared into the darkness . . . The frightened yells of my escort told me that a *ghoul*, a wicked demon of the desert, had seized the camel; the flintlocks came down and several shots were fired to chase away the evil spirit.

Notwithstanding this bizarre happening, Vischer liked camels (p.105–6):

> I have often admired my camels for their calm, well-bred behaviour, so in keeping with the desert . . . Once loaded up, they go straight on without any trouble . . . The camel I rode from Bilma to Bornu soon became attached to me, like a dog, and would follow me about everywhere. . . Many a night, to guard against the piercing wind, I slept alongside my good warm animal, which took me to his bosom like a mother.

Towards the end of the journey they passed through great waves of sand. As Vischer recalled (p.277):

> We usually travelled at night in order to spare our water, and as much as possible to save the camels, who began to fail in an alarming manner. We soon had to face the prospect of abandoning parts of our loads. When a camel is exhausted, it shows the first signs of distress by lagging behind and moving its long legs in a stiff and jerky way, very different from the easy gait with which it usually swings along. It is fatal, then, to let the animal kneel down, for the exertion of rising is always great, especially when it is loaded; finally, when the good creature feels its force at an end, it seems to become heart-broken and sinks down on the sand. Nothing will ever get it up again; with an expression almost human in its resignation and misery, it will watch with its large sad eyes the caravan slowly passing by and disappear over the next dune.

Vischer was pleased to reach his beloved Bornu, but had been much moved by the Sahara (p.293):

> The desert has left an impression on my soul which nothing will ever efface. I have entered it frivolously.…I left it as one stunned, crushed by the deadly majesty I had seen too closely.

Vischer's subsequent life was not without incident. In 1910 he was appointed the first director of education in Northern Nigeria. Vischer devoted his life to the progress of education in Africa. With his cosmopolitan background and strong sympathy with African culture, he made an important contribution to the evolution of British educational policy. Between 1915 and 1919 he served in military intelligence, and from May 1917 was based in Bern with the rank of major. He ran an SIS (Secret Intelligence Service) network there and in Spain. With the onset of the Second World War, Vischer again served in intelligence, working for the Department of Propaganda to Enemy Countries, and served as its liaison officer with MI6. He was made a CMG in 1936, and knighted in 1941. He died in 1945. His obituary in *Nature* (Tomlinson 1945) remarked that in his presence:

> All prejudice, British or African, melted like snow in the midday sun; and the reason was not far to seek. The Hausa people, quick at all times to detect the idiosyncrasies of their alien overlords, proclaimed the secret

of his influence in two words. Throughout the country he was known as Dan Hausa ('Son of Hausa'). It meant that in him the Hausas recognized, not only a man who possessed an extraordinary command of their delightful and expressive language, but .one whose affection for them was equalled by his insight into the innermost recesses of the African soul.

## NOWELL BARNARD DE LANCEY FORTH

Nowell Barnard de Lancey Forth (1879–1933) (**Figure 2.33**), was born in 1879 at Dwaroon, near Warrnambool, Victoria, son of De Lancey Forth, farmer, and his wife Annie Thompson. He was an ancestor of Rupert Murdoch the newspaper proprietor. After attending Geelong Grammar School he enlisted as a private in the 4th (Queensland Imperial Bushmen) Contingent and embarked for the South African war in May 1900. Promoted sergeant in June and twice recommended for a commission, he was gazetted to the 3rd Manchester Regiment, British Army. He saw action in the Orange River

**Figure 2.33** Lt Colonel Nowell Barnard de Lamcey Forth, DSO and bar, MC, while a major in the Imperial Camel Corps Brigade. Portrait by J. St Helier, 1917.
*Source: Forth 1991.*

Colony, and near Pretoria in the Transvaal. After this service he was invalided to Australia. In July 1901 he was promoted to lieutenant in the Manchester Regiment and in 1901–06 served with the British regular forces. Seconded to the Egyptian Army in March 1907, he spent the next nine and a half years as a member of the Sudan Camel Corps. Promoted to captain in the Manchester Regiment in 1912, he was mentioned in dispatches, attained his majority and was awarded the Military Cross for his service in the Nuba Mountains. In 1917 he commanded the 3rd Battalion, Imperial Camel Corps. Then, by now an acting lieutenant-colonel, he was transferred in June 1918 to command the 7th Battalion, Essex Regiment, also in Palestine. As well as his M.C., D.S.O. and Bar, he also received three mentions in dispatches for his war service. Promoted temporary lieutenant-colonel in December, he was stationed until 1920 at Sollum, on the Egyptian-Cyrenaica frontier, in command of the Western Desert Forces. In April 1920 he was employed by the Egyptian Government as commander of the Camel Corps and Light Car Patrols, Frontiers Districts Administration. Stationed mainly in Cairo he remained in command of this unit until 1924. During that time he sought Zerzura (unsuccessfully) on camel treks conducted between Siwa, Ammonite Hill and Regenfeld. In 1924 he contracted an illness which for many years made him an invalid, and so his exploring career was cut short. A great admirer of Rosita Forbes, and never a great exponent of the motor car for exploration, he died in Alexandria in Egypt in 1933. His obituary by Harding King in the *Geographical Journal* of that year remarked that he was 'One of those men of whom the public hears little, since their work lies on the fringe of civilization and they are not given to advertisement, but who from long experience, and a natural aptitude for their work, are able to outwit the wild native tribesmen with whom they have to deal and to beat them on their own ground'.

Forth's time in the Camel Corps has been described by his cousin, Nevill de Rouen Forth (*A Fighting Colonel of Camel Corps*, 1991). In 1917 in Sinai, in the battles against the Turks, camels were used for evacuating the wounded (p.97):

> During all the actions the stretcher bearers displayed the greatest courage as they moved calmly across the open country in front of the enemy position, picking up the fallen under very heavy fire and removing all wounded from the field, carrying men to the ambulances. Some were drawn by camels and others were sand carts drawn by mules. Cacolets were also used throughout the battles in which the Imperial Camel Corps fought. Cacolets were stretchers, one on each side of the camel

> pack saddle. For sitting cases there was a chair on either side. Both types can only be described as fiendish contraptions, tossing and rolling like a rudderless yacht in a hurricane. They were exhausting for the wounded who suffered great pain and discomfort, sometimes travelling like this for several days across waterless desert. Limbs and lives were lost through gangrene.

His search for Zerzura in the 1920s was written up in the *Geographical Journal* in 1930. In 1921, accompanied by Colonel Macdonnell, the then Governor of the Western Desert of Egypt, he penetrated with a Camel Corps patrol some 160 km into the unknown dune area south of Siwa. He found this journey was so interesting that he determined to try and repeat it. In 1924 he made a second journey, in company with Captain Hatton, M.C, of the Frontier Districts Administration.

On this second journey he penetrated to a point about 320 km due south of Siwa. Using this and other journeys, he estimated distances by the pace of the camels, and in this way prismatic compass traverses were compiled.

**Figure 2.34** Forth spent much time fighting in Sinai, and in 1922 he visited St Catherine's monastery and Gebel Musa with Lord Allenby. *Source: Author.*

Between these journeys, in the winter of 1922–1923, he investigated the country west of the Dakhla-Abu Mungar escarpment westward to a point near Rohlfs' Regenfeld in the south and his Ammonite Hill in the north. In 1922, Forth was given a major task. As Officer Commander of the FDA Camel Corps, he had to accompany Lord Allenby to Sinai, and in particular to St Catherine's monastery and Gebel Musa (**Figure 2.34**). They travelled on horseback but with a camel escort.

On the basis of these journeys he offered advice on camels (p.58):

> I would advise a reserve of camels and water double that one would take on any other journey of the same length in ordinary desert country. With regard to camels, breeding counts, and I have always found well-bred seasoned riding-camels from the Sudan, not too short in the leg, with strong hind quarters, well ribbed up, and with good chests, the best animals for the heavy work to be met in the dunes. Their load, apart from the man they carry, should be light.

He kept up his interest in Zerzura to the end, but while staying with Wiffy Jennings-Bramly at Borg-el-Arab he contracted pneumonia. He died in an Alexandria hospital, his hand held by his old friend, Wiffy.

## ARTHUR WEIGALL

Egypt's Eastern Desert lies between the Nile and the Red Sea. It was a little studied area, though it had been visited on camel by a great French engineer and artist, Louis-Maurice-Adolphe Linant de Bellefonds (1799–1883) in the 1820s. He wrote much about the Bisharin nomads and their camels (de Bellefonds 1868) (**Figure 2.35**). He also rode to Meroe, Petra, Siwa and Abu Simbel. In the 1890s, the conservationst Edward Buxton and his daughters went by camel from Qena on the Nile across to the Red Sea (Buxton 1894).

Later, it was in the Eastern Desert that Arthur Weigall (1880–1934), a British Archaeologist and Egyptologist, made notable explorations. Born in 1880, he never knew his father, an army officer who died in the same year while on active service in Afghanistan. His mother was a revivalist missionary who worked to change the lives of destitute prostitutes. He attended Wellington College, New College, Oxford, and Leipzig before finding work with Flinders Petrie, a founder of scientific archaeology and meticulous excavation,

**Figure 2.35** Bisharin nomads in the Eastern Desert, as drawn by de Bellefonds.
*Source: de Bellefonds, 1868.*

first at University College London and then at Abydos in Egypt. Subsequently, aged only 25, he was appointed to replace Howard Carter, of Tutankamun fame, as Chief Inspector of Antiquities for Upper Egypt at Luxor. He threw himself with enormous energy into trying to thwart the export of antiquities and to conserve what was left of Egypt's heritage. He was critical of Carter's efforts in this area, but himself did a great deal to conserve, *inter alia*, some of the monuments that were threatened by the construction and heightening of the original Aswan Dam.

In 1909 Weigall published his account of *Travels in the Upper Egyptian Deserts*, which was based on a series of expeditions that he had conducted. He was captivated by the Eastern Desert, praising it for its clear atmosphere, the lack of evil odours, the purity of the wind, the richness of archaeological sites and ancient mines, the scatterings of prehistoric worked stones, and the amazing night skies. Weigall travelled by camel. He remarked (p.25) that 'With a steady steed and a good saddle there are few means of locomotion so enjoyable as camel-riding'. He reckoned one could cover some thirty miles a day with ease. Weigall's caravan consisted of himself, a native assistant, a servant and a guide, together with a baggage-train of a dozen camels and men and a couple of guards. Of the guide he wrote (p.115):

> The guide was a picturesque, ragged old man, whose face was wizen and wrinkled by the glare of the desert. His camel was decked with swinging tassels of black and yellow, and across his saddle there was slung a gun at least seven feet long, while at his side there hung a broad-bladed sword in an old red-leather case. In his belt there were two knives, and in his hand he carried a stout bludgeon, something in the form of a hockey-stick.

However, even in 1909 he recognised that future travel might well be by motor car, that a journey from the Nile to the Red Sea might in due course be accomplished in a morning and that more and more travellers would venture out beyond 'the little garden of the Nile'.

Being an archaeologist, Weigall was interested in the history of the camel in Egypt (p.24):

> In ancient Egyptian times the camel was regarded as a curious beast from a far country, and was seldom, if ever put to any use in Egypt. Only three or four representations of it are known, and it never occurs amongst any of the animals depicted upon the walls of the tombs, although bears, elephants, giraffes and other foreign and rare creatures, are there shown. It was an Asiatic animal, and was not introduced into Egypt as an agent of transportation until the days of the ubiquitous Romans. Donkeys, oxen and human beings were alone used in Pharaonic days for transporting the necessities of the labourers and the produce of their work; and probably the officials were carried to and fro in sedan- chairs.

Weigall, glad to get away from professional conflicts in Luxor, was a devotee of the wilderness (p.150–1):

> Here there are no cares, for there are no posts or newspapers; here there is no fretfulness, for one is taking almost continual exercise; here there is no irritation, for man, the arch-irritant, is absent; here there is no debility and fag, for one is drinking in renewed strength from the strong conditions around . . . The desert is the breathing-space of the world, and therein one truly breathes and lives.

Weigall wrote extensively on the archaeology of Egypt, including studies of Cleopatra and the Pharaoh, Akhenaten. He remained in Luxor until 1911 and, following a breakdown, largely brought about by professional rivalries, left Egypt in 1914. Thence he changed his career almost completely, transferring from the Eastern Desert to the West End. In London he became a set-designer, lyricist, novelist, film critic and journalist, before returning to Egypt in the 1920s to cover the opening of the tomb of Tutankhamun for *The Daily Mail*. He died in 1934 aged only 53 after a quick and painful illness. Some suggested that he was the victim of the boy king's curse.

## GEORGE MURRAY

George William Murray (1885–1966) (**Figure 2.36**) was a key figure in the exploration of the Libyan Desert. A Scotsman from Aberdeen, he joined the Survey of Egypt in 1907, succeeded John Ball as Director of Desert Surveys in 1932–1937, and became director of the Topographical Survey between 1937 and 1947. He played a major role in the late 1920s in surveying the newly discovered Qattara Depression and the country between the Depression and the Mediterranean. Murray was awarded the RGS Founders' Medal in 1936 (Bagnold, 1966). He also operated in the Eastern Desert of Egypt, and during his surveys he discovered ancient Egyptian diorite quarries and carried out the first archaeological investigations of the medieval Red Sea port of Aidhab (1926). Indeed, archaeology was his passion, and he contributed many papers on Roman and other archaeological sites. He also scaled some of the highest mountains of the Eastern Desert, being the first European to do so (Murray 1947). He also edited an English-Nubian dictionary.

**Figure 2.36** George William Murray.
*Source: Murray 1967, frontispiece.*

Murray, 'Desert Murray', a friend of the Arabian explorer Freya Stark, was a lover of the desert who participated in several expeditions with his wife Edith and two large Airedale dogs named King Odd and Prince Worricow. Murray himself was of very substantial size and stood in marked contrast to the diminutive Ball. There is a splendid photo of him in the frontispiece of his autobiography *Dare Me to the Desert*, wearing voluminous shorts and, horror of horrors, calf-length dark socks. During the Second World War the maps of Murray and his colleagues were to prove to be of inestimable value to Ralph Bagnold. Bagnold regarded Murray and his wife as 'close and lasting friends'. They enjoyed mountaineering together. Murray was also a friend of another explorer of Sinai and the Libyan Desert, Wilfrid Jennings-Bramly.

Murray used camels in his survey work, though in later years he adopted the motor car. *In Dare Me to the Desert*, he recounts the first major trip he seems to have made in the area between the Nile and the Red Sea. He described his men (p.19):

> Ragged and half-naked though they were, the 'Abada camel-drivers and porters were a delight to the eye. There were about a dozen jaunty youths and young men in the party without an ounce of fat on their bodies, though one or two of them carried a good deal of it in their shocks of hair . . . All were erect and muscular; all chattered away and guffawed at the top of their voices.

In 1926 he contributed a very short note to the *Royal Geographical Journal* on the ability of camels to go without water if good grazing was available:

> Copious rains fell in this district, Gebel Elba in the Eastern Desert of Egypt, lat. 22 0 N., last November, and the vegetation that sprang up has been kept green by dew and further showers. The camels of my survey expedition drank last near Mersa Shab on November 22, and my sheikh el hamla does not propose to water them till about the end of Ramadan, April 14, perhaps. That is to say, they have already gone 103 days, and may go another 40 days, without water. They have moved my camp by very easy stages 440 kilometres in the 103 days. As to whether grazing, not working, camels could go six months in the cool of the year on green grazing neither my guides nor myself entertain any doubt whatever. But in hot weather it is a very different matter. I suppose Hassanein Ahmed Bey's march in ten days from Owenat to Erdi, 450 kilometres, in May represents about the best that could be done without either water or grazing.

However, Murray has significance for the study of camels because in 1935 he wrote a major study called *Sons of Ishmael: A Study of the Egyptian Bedouin.* It contains a whole chapter devoted to Bedouin camels. He recognised that camels and the Bedouin were inseparable (p. 104):

> To this ungainly animal the Bedouin owes nearly everything, and he repays this with as much affection as his rather material culture can muster. They love their camels . . . An old Arab once said to Bramly 'You cannot imagine my joy at being kept awake all night by the bellowing of my camel-calves!' and again 'A man of property is he who sitting on the ground, sees his horizon limited in every direction by the bellies of his camels!'.

Murray also discusses such diverse issues as the use of camels for pulling carts and wagons full of mining machinery and railway tracks, the treatment of their foot pads with henna, predation of foals by jackals or foxes, their susceptibility to poisonous flies in Middle and Lower Egypt, their dentition, the virtues of different breeds, the nature of saddles, the possibility of obtaining water from a dead camel's stomach, their wonderful sense of direction in pursuit of water, and the superstition, unfounded in Murray's view, that camels cannot breed unaided. For the most part he suggested that except when rutting, the camel is mild and docile, though he did recount one unfortunate incident (p. 112):

> Once I heard of a youth being killed by his camels at Bir Mab'uq near Suez. He had been wearing a newly-killed jackal's skin over his shoulders, and the smell of this so enraged the milch camels that they trampled him to death with their knees.

Apropos of rutting, Murray recounted that one stallion camel ran away from his camp on the Red Sea to meet up with his beloved, and was caught next day near Qena on the Nile – a distance of 240 km.

In 1951, Murray was thrown out of Egypt. He died in Aberdeen, aged 80.

## ANGUS BUCHANAN

Angus Buchanan (1886–1954) (**Figure 2.37**) was born in Kirkwall on the Orkney Islands. In 1915, he joined the 25th Battalion of the Royal Fusiliers, also known as the Frontiersmen, and served until October 1917 in East Africa until he was wounded whilst fighting and was sent home. He was awarded the

**Figure 2.37** Buchanan on his camel.
*Source: Buchanan 1926.*

Military Cross. Buchanan describes a march of more than 1,370 km from British East Africa (present-day Kenya) into German East Africa (present-day Tanzania).

His first trip, to the Aïr Mountains of the Sahara, was in 1919 with a 2250-km expedition financed by the 2nd Baron Rothschild, who was keen to find new species of animals in an unexplored part of Nigeria. He met up with the Tuaregs (p. 232) 'the finest travellers and camel-riders in the world'. On returning from this trip, Buchanan brought back over 140 mammals that Rothschild gifted to the Natural History Museum. Eighteen of these were newly discovered species. Buchanan followed-up the trip with a book titled *Exploration of Aïr: Out of the World North of Nigeria* (1921). In the introduction, Rothschild indicated that he was hugely impressed (p.xxiii):

> Considering the long journey by camel and the fact that Captain Buchanan was working absolutely single-handed, the collection of over 1,100 Birds and Mammals and over 2,000 Lepidoptera, in a region notorious for its paucity, both of species and of individuals, is a remarkable achievement, and proves him to be a most efficient explorer and naturalist.

Buchanan was keen on his camels (p.43):

> The awkward and somewhat wooden appearance of camels does not lead one to associate much intelligence with them, but to think so is a mistake, and if one desires to have a really good hunting camel, I know of no better method to secure it than to select a good-natured beast from the rank and file, and hand feed it with tit-bits of vegetation, and pet it when mounting and dismounting, and let no one else saddle it or ride it, and before long you will be astonished to find that you have won a queer pet and a useful and obedient comrade.

In March 1922, Buchanan set out with the ambitious aim of crossing the Sahara from South to North. Among those with him for part of the journey was Francis Rodd, who became a considerable expert on Aïr and the Tuareg – 'the people of the veil'. Parts of the expedition were filmed by a Canadian, T.A. Glover. The film, shown in 1924, was titled *Crossing the Great Sahara* and its introduction claimed it was 'the First Photographic Record Ever Secured of the Life, Vastness and Mystery of the World's Greatest Desert'. In 1926 Buchanan produced his book *Sahara*, with its charming art deco cover.

Buchanan and Glover, travelled from Kano in Nigeria to Touggourt (Algeria), a journey of over 5600 km. It lasted fifteen months. Some idea of the hardships which they encountered may be gathered from the fact that, while they started with a caravan of 36 camels and 15 Africans, they finished with only one camel and two Africans.

Early in the journey, one of the camels, Mizobe, expired (p.23) and Buchanan displays his compassion for ailing camels:

> We find that he has collapsed wearily on the sand and does not want to move. He is far through, but we cannot camp and wait beside him. So in a little time he is persuaded to rise to his feet again; and the caravan moves slowly on. But it is not very long before the poor old fellow gives up again, for his is a losing fight in the full heat of the midday sun. We try for a little to encourage him to get up, but to no avail. He is past further struggle. I order the caravan to move onward, and remain behind with Elatu, beside the prostrate animal, for I cannot leave the poor brute to die a slow, lingering death, with the agony of pitiless surroundings holding finality immediately before his eyes. When the caravan is distant there is a single revolver shot and we are one less in our band.

Perhaps the most interesting portion of Buchanan's book is his lengthy description of the great *Taralum* caravans associated with the Bilma salt trade (p.71 et seq.):

> We had ridden in upon a camp of outstanding proportions and unique picturesqueness. Before us stood thousands of camels, not a hundred or two, which would have been amazing enough, but, literally thousands; and the spectacle was never one to be forgotten. Where the ruins of the old forsaken village of Tabello squat dolefully on the banks of the river of that name, the great caravan had already congregated in part, and was still in

> process of expansion. As far as the eye could see camps were settled on the banks and on the sand of the river-bed – camps full of packsaddles, water-skins, bundles of coarse hay-fodder and bundles of firewood; all in readiness for the long desert journey. About the camps, among the camels, picturesque camel-men moved gracefully, or recline upon the sand – athletic-looking men. . . . They are chiefly Tuaregs . . . The whole concourse was gathered from far and wide to this appointed rendez-vous: from Kano, Katsina, and Sokoto, in Northern Nigeria; from Gourè and Zinder and other towns in Damagarim and from many quarters in Damergou and Air.

Buchanan marvelled (p.74) at how the camels, given the fact that their 'lips and noses are soft as velvet and sensitive in the extreme', were able to eat sharp-thorned acacias. He also reported that the caravan was protected by a force of Meharists supplied by the French administration. Every member of the caravan itself was armed with a rifle, sword or lance, and some carried a shield made of oryx hide. He remarked that the date of departure was an event in the Sahara that was as 'notable as Christmas Day in civilised countries'. On p.83 he discussed the numbers in the caravan:

> In the open desert the Taralum made an astonishing array. The space that the 7,000 camels occupied on the march is almost past belief. From a situation in the centre of the caravan one viewed neither the head of the cavalcade nor the tail. So far as eye could see, out in front, or back in the rear, the marching army diminished until vanishing lines met the horizon, dark specks on the light sand, looking like mere swarms of flies on the carpet of the world.

On reaching the end of the journey, Buchanan (**Figure 2.38**) once again displayed his compassion for camels (p.262):

> About midday I parted from my camel, Feri n'Gashi, to whom I was tremendously attached, for he had faithfully carried me throughout the journey. He seemed to understand that the end had come, and it was a strange, sad-eyed farewell between master and dumb friend . . . He looked well, considering all he had gone through, and I sent him away to enjoy a well earned rest, having arranged with Captain Belvalette that he should return to Ouargla and be cared for so long as he lived. I had no inkling of the rapid sequel. The rest he was to have was of another order,

**Figure 2.38** Buchanan with Ali, Sakari, and the faithful but fated Feri n'Gashi.
*Source: A. Buchanan,* Sahara, *1926.*

> for in the afternoon Ali came running to me in consternation to tell me that Feri n'Gashi was dead.

Little seems to be known about the rest of Buchanan's life. During the 1920s and 1930s he published two novels: *The City of the Seven Palms* (1927) and *Gain: A Tale of the Canadian Northland* (1932). During World War Two he fought again, serving in Abyssinia until he was injured in 1943. He died in 1954 in Edinburgh.

## AHMED HASSANEIN

Ahmed Hassanein (1889–1946) (**Figure 2.39**), a member of the Turkish ruling class in Egypt, captained the Egyptian team at the Brussels Olympic Games in 1920. He enjoyed the patronage of King Fouad I and his son, Farouk and became an influential figure in palace affairs. Born in Cairo in 1889, where his father was a distinguished scholar at the thousand-year-old mosque of

**Figure 2.39** Ahmed Hassanein. *Source:* National Geographic, *September 1924.*

al-Azhar, he was educated at Balliol, an Oxford college that was at the time regarded by both itself and outsiders as an intellectual power house. There he was a friend of Francis Rodd, who himself became a noted Saharan explorer, and together they served during the First World War on the British mission to the Senussis in the Libyan Desert. It was then that they conceived the idea of visiting the so-called secret oasis of Kufra, but in the event Rodd dropped out and was replaced by Rosita Forbes (1890–1967). Soon thereafter, however, he joined Angus Buchanan in the Aïr region. Ms Forbes, who withstood the hardships of her journey with great stamina and laudable fortitude, is often accused of magnifying her own role and downplaying that of her Egyptian colleague when writing her account of the expedition.

During the trip this divorced woman, strikingly handsome, with dark hair and eyes and known in London for her huge Ascot hats, high-heeled shoes and sophisticated make-up, dressed as a Muslim woman and posed as a relative of Hassanein. She took the name of Khadija and invented a Circassian mother to account for her imperfect Arabic (**Figure 2.41**). It is said that she tried, unsuccessfully, to seduce him but he resisted her advances. Often called 'Sita', she was born in England in 1890. Her journey with Hassanein in 1920–1921 was her greatest feat of true exploration, but she later travelled very extensively, partly as the backdrop for a string of novels that she produced. She went to Morocco, Abyssinia, Persia, Saudi Arabia, Yemen, Afghanistan and Central Asia. She died in 1967.

Rosita described their joint journey in *The Secret of the Sahara; Kufara*, in 1921. She dedicated it to Hassanein 'In memory of hours grave and gay, battles desperate or humorous, of success and failure in the Libyan deserts'. She described their camel trek from Jedabia, south of Benghazi, to Aujela, Jalo, Buseima, and Kufara, and their return via Jaghabub to Siwa and Cairo.

They endured a dreadful Gibli (sandstorm) after leaving Jalo (p.116):

> The air became a thick sheet of sand. Sun and direction were blotted out. Screaming gusts stung our faces and blinded us. It was the most extraordinary sight, for one minute camels and figures would be blotted out in a whirling white fog; then a head or a wildly- gyrating blanket or a portion of a labouring camel would appear for an instant through the shroud and then vanish again in the smarting dusk.

Notwithstanding such events, on one occasion Rosita managed to sleep on a camel (**Figure 2.40**).

Rosita was the first non-Muslim woman to enter Kufra. She was received with clamour in London and was in demand as a speaker at lunches and dinners, a challenge she took up with verve. She spoke to a full house at the RGS and was even welcomed to Buckingham Palace, where she sat on a gilded sofa between King George V and Queen Mary, showing them her map of the Libyan Desert.

**Figure 2.40** Rosita Forbes asleep on a camel (the one second from the left).
*Source: Forbes 1921.*

Some people were less than glowing in their appreciation of this expedition. For example, D. Newbold wrote slightly maliciously (1924, 103):

> This journey, though a very gallant undertaking and considerably advertised in the English and Continental Press did not actually add much to the information gained by Gerhard Rohlfs 40 years previously; and Mrs. McGrath's resulting book largely proved, in spite of its vivacious style and excellent photographs, a disappointment to those who had hoped for some scientific results. It was, perhaps, not to be expected that an expedition which had perforce to be ill-equipped and had to encounter such extraordinary physical difficulties would be able to bring back any detailed data but the lack of observations on such subjects as geology, ethnology, etc., and, not the least the rather unreliable topographical results, left plenty of scope for subsequent travellers.

In the first half of 1923, Hassanein made his great camel trek from Sollum on the Mediterranean to El Obeid in Sudan (**Figure 2.41**). He did not take Rosita with him and he rectified some of the shortcomings that Newbold had noticed. As he wrote in the *Geographical Journal* (1924, 274):

> On the previous trip we had taken no scientific instruments, except an aneroid barometer and a prismatic compass. It was not possible, therefore, to make exact scientific observations. In fact we brought back merely the notes for a simple compass traverse of the route, which I had made as best I could with the meagre means at my command. I was determined then that my new expedition should be equipped with the necessary scientific instruments for surveying the country traversed, and if possible contribute to the geographic and topographic knowledge of the Libyan Desert.

He covered c.3300 kilometres over seven months and 23 days. It was the climax of pre-car desert explorations. For this he was awarded the Founder's Medal by the RGS in 1924. The trip established the true location of some major landmarks of the Western Desert, including Kufra, Arkenu and Uweinat. Hassanein found many rock pictures at Uweinat and these excited many later explorers to visit the area.

John Ball, head of the Egyptian Government's Desert Survey Department (*Geographical Journal* 64, 385) was particularly impressed by the surveying skills Hassanein had learnt, as before setting out from Cairo he had applied himself

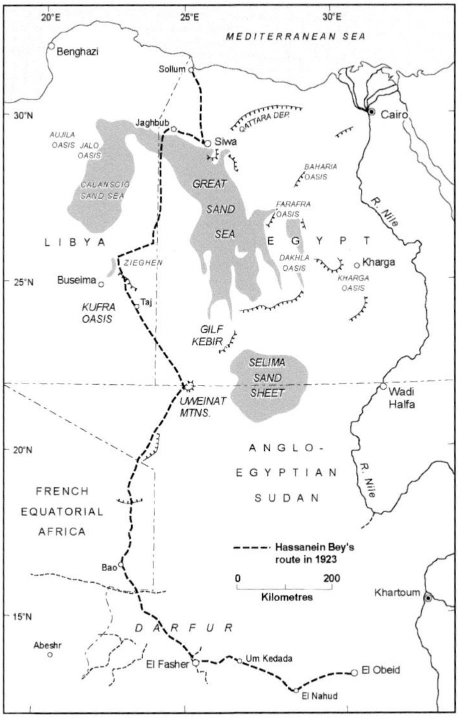


**Figure 2.41** Ahmed Hassanein's great journey of 1923. *Source: Author.*

assiduously for some weeks to acquiring expertise in the use of the theodolite, and instruction in the particular methods of reconnaissance-survey best adapted for exploration of the kind on which he was to embark.

Hassanein wrote about his journey in *The Lost Oases* (1925). He had good things to say about the Bedu and their camels (p.27):

> I never cease to marvel at the Beduin serenity and courage, which nothing disturbs. In desert travel there are three elements – camels, water supply, the guide. Camels, the best of them, and for no apparent reason, give in, as it happened when I left Kufra and one of my best camels died on the second night; while, on the other hand, the weakest camel of the caravan, which left Kufra tottering under its load, went through the whole trip, about 950 miles, and arrived tottering at El Fasher. 'God will protect it', said its Beduin owner when rebuked for bringing such a sorry animals, and in truth God did protect it.

One Beduin said (p.124) 'He who will not risk his life for his camel does not deserve to have it.' On their longer stretches, with no grazing on the way, they exclusively fed their camels on dried dates (p.103). Their caravan consisted of thirty-nine camels, twenty-one men, a horse and a dog (p.116). After twelve or thirteen hours of walking they came to the end of the day's trek (p.125):

> '*Eddar ya ayan* – home for you who are weary' is shouted by the guide and repeated by every man in the caravan. Then the men collect the camels and divide them, the water camels here, those carrying tents over there, the camels with luggage for the barricade yonder. The camels are *barrakked* – kneel with grunts of satisfaction to have their loads removed.

Hassanein prized some aspects of the camel (p.131):

> The qualities of a camel are seldom, if ever, appreciated on a slight acquaintance. The camel is as clever as a horse, if not more clever, and in some ways is more human. 'Patient as a camel' is an Arab saying and a very true one. If you ill-treat a camel he will never forget it, but he will not attack you on the spot. He will wait, and if you repeat the offence again and again, he makes up his mind to get his revenge . . . He watches his chance until you and he are alone and then he goes for you . . .

He also noted how camels tried to keep in the safety of the caravan, and how they have a memory for wells they had visited long ago. He appreciated the fact that in winter a well-trained camel can survive for a fortnight without water, and in the summer up to twelve days. He also recognised that 'the camel can become a very affectionate beast and devoted to its master' (p.134).

In the *Geographical Journal* of 1924, he described the nature of camel travel (p.285):

> I had one camel fitted up with a tent over its back so that one may lie and take one's ease when tired of walking. Ahmed called it 'the Club'. One day at the lunch hour Abdullahi demanded where I was and whether I had had my portion of bread and dates or not, and Ahmed replied, with a twinkling eye in an otherwise grave face, 'The Bey is lunching at the Club today'.
>
> It is quite possible to have a good nap on the camel's back, and an occasional ride is not to be scorned. But generally one walks, for the camel's pace of 2½ miles an hour is an easy one to keep up with, and riding is often more tiring than walking.

He concluded his trek at El Obeid. From there, Hassanein took a train to Khartoum and thence to Cairo. In America, which was gripped by the Rudolf Valentino film, *The Sheik,* he became a celebrity. In Egypt, his explorations were regarded as a patriotic achievement. He was feted at a ceremony in the Cairo Opera House, where King Fuad conferred on him the title of *bey*. In 1927 the British made him a Knight Commander of the Victorian Order. They also inadvertently caused his death, for in February 1946 he died in a freak motor accident when a British truck skidded and smashed into his car on a rain-swept Cairo bridge.

## LUDIVICO MARIANO NESBITT

**Figure 2.42** L.M. Nesbitt. *Source: photo by Eva Barrett, 1930, taken from Nesbitt 1930.*

Ludovico (Lewis) Mariano Nesbitt (1891–1935) (**Figure 2.42**), a British explorer and mining engineer, was the first European to traverse the Danakil Desert of eastern Africa and survive. Born to a Scottish and Italian family in 1891, he packed a great of travel into his life before dying at the age of 44 when the Dutch airliner on which he was flying crashed in the Swiss Alps.

Trained as a mining engineer in Rome, the Cranborne School of Mines in Cornwall and in the Witwatersrand in South Africa, he began his first work in the South African gold mines. Subsequently, he went to Cuba and the Llanos of Venezuela. In 1927 he travelled from Sudan, where he had been prospecting, to Abyssinia. His ambition was to explore Danakil, the stretch of torrid terrain situated between the Ethiopian Plateau and the Red Sea.

Before his expedition, there had been three other journeys to the region – those of Werner Munzinger in 1875, Giuseppe Maria Giuletti in 1881 and Gustavo Bianchi in 1884. All these expeditions ended in disaster and no European member of any one of them ever returned. Notwithstanding this, Nesbitt, with two Italian companions, Giuseppe Rosina ('a wonderful packer' of the camels) and Tullio Pastori, set off on an 1300-km march lasting for three and a half months. Pastori was a self-confessed spy for Mussolini, who was to invade Abyssinia in 1935. Travelling from south to north, much of it down the Awash Valley, they made compass traverses covering over 50,000 square km. They had with them fifteen 'natives' of Abyssinia and Eritrea, and a caravan consisting of twenty-five camels (which he branded with a home-made branding iron) and four mules. Three of their 'native attendants' were killed by the Danakils (whose reputation for ferocity was legendary). Ten camels and three mules died of thirst, starvation and fatigue.

The journey was made in secret, without the permission of the Abyssinian Government and without informing the British or Italian legations. They feared official opposition. They set out from a location on the railway line called Awash Station. They soon encountered the Madima tribe (*Hell-hole of Creation*, p.135):

> Many of these Madima men had bracelets, worn between the elbow and the shoulder. These ornaments indicated that the wearer had slain a man or men, each bracelet representing one victim. Every grown man also had a hole pierced in the lobe of each ear, a sign that a year had elapsed since he had killed his first victim. The younger ones wore instead a feather stuck in their long woolly hair, to show that though they had not been murderers for a whole year, they had nevertheless already been successfully engaged in the honourable pastime. The feather is worn during the twelve months immediately following the first kill. All these men looked savage and bloodthirsty.

He was not fond of the Danakils (p.164):

> . . . there is no more slothful race of men than the Danakil. They are indeed more like wild animals than men, sleeping on the ground, living almost exclusively on raw meat and milk. No work spoils the elegance of their supple feline bodies . . . The lines of their faces are clean-cut, angular, chiselled, and somewhat effeminate.

They encountered camel caravans transporting salt from French Somaliland (now Djibouti) to the Ethiopian Plateau. They were also beset by ticks, wriggle-worms and pestilential horseflies. One of the worst problems was the extreme heat, with temperatures regularly exceeding 60°C (p.249):

> As soon as the sun rose, the heat became torrid. We were soon in motion on the burning desert, which was here intersected by walls of basalt. These stood out in sharp lines of metallic blackness against the white dazzling plain. Not a sign of life was visible; there was no animate thing besides ourselves in the whole glistening wilderness.

At one water hole they encountered nature at its most cruel (p.270):

> We were awakened at dawn by the cries of partridges, and the hissing of snakes . . . At sunset, the reptiles creep out of their hiding-places,

> and prepare to prey upon the birds as they alight to drink. At night, the despairing cries of the victims, and the mad flapping of their wings, are continually heard.

Nesbitt admired the way that the camels could eat dry thorn bushes (p.142):

> They would close their jaws on the rough twigs, and then with a sideways pull, strip them of the bark and thorns and of any buds which grew on them. Sometimes, large thorns would stick in their gums or lips, and then they would show visible appreciation when one removed them. The longer you associate with your animals, the more human they seem to become, especially when the association is as close as ours was on this expedition.

The Nesbitt party suffered greatly from the absence of water as they traversed the volcanic terrain, and one of their camels expired (p.336):

> . . . one of our camels cast himself down, and had to be abandoned. . . . It was one of the last we had purchased at the outset, at Awash Station. It fell into one the subsided domes, and could not rise again. After giving the faithful beast a farewell caress, I took a rifle from Wold Jesus and shot it.

Very soon thereafter they reached the saline Egogi Lake (Afrera), the first Europeans ever to do so, and a group of five smoking volcanic cones at Hertale (Erta Ale), the greatest reward for the privations they had undergone. They also saw Lake Assale, with its huge salt deposits, and one of the lowest spots on the face of the Earth. Eventually they reached the Red Sea port of Massawa, and their journey was over. Some years later another extraordinary explorer, Wilfred Thesiger, was also to brave Danakil. Thesiger (*Geographical Journal*, 1934, 527), though critical of some of Nisbett's statements and expressing misgivings about the scientific exactitude of his observations, admired his 'vivid and distinctive prose', but most of all was struck by his 'unshakeable patience, courage, and determination'.

The year after Nesbitt's journey another Italian, Baron Raimondo Franchetti (1889–1935), carried out further explorations in Danakil, travelling from east to west and also using camels. Like Nesbitt, in 1935 he died in an aeroplane crash.

# SIR CHARLES DALRYMPLE BELGRAVE

Sir Charles Dalrymple Belgrave (1894–1969), known generally as Carol, was educated at Bedford School and Lincoln College, Oxford. During the First World War he served in the Imperial Camel Corps in Sudan, Egypt and Palestine. After the war he was seconded to the Egyptian Government to help the Frontier Districts Administration in the Siwa Oasis.

Belgrave wrote an engaging book, *Siwa, the Oasis of Jupiter Ammon*, which was published in 1923. It gives a good impression of the life of an officer in the Camel Corps, but also tells us much about Siwa, an oasis in the interior of the Libyan Desert in western Egypt. It contains some lovely coloured images of the oasis, which although unattributed, were probably by the author himself (**Figure 2.43**). He was less than flattering about the men of the oasis (p.149–50):

> The Siwans are typically Oriental. They are hospitable, dishonest, lazy, picturesque, ignorant, superstitious, cheerful, cunning, easily moved to joy or anger, fond of intrigue and ultra conservative. They are not immoral, they simply have no morals. The men are notoriously degenerate and resemble in their habits the Pathans of India. They seem to consider that every vice and indulgence is lawful.

On p.38 he discusses the Frontier Districts Administration (FDA) Camel Corps, which he commanded:

> The F.D.A. Camel Corps was originally formed of Sudanese men from the Coastguard Camel Corps, with a large proportion of 'yellow bellies' (Egyptians) who were gradually weeded out and replaced by Sudanese and Sudan Arabs, who were enlisted on the borders of Egypt, as the Sudan Government does not allow recruiting inside its territories except for the Egyptian Army. The F.D.A. Camel Corps is supposed to consist entirely of Sudanese, but a certain number of the men who were enlisted in the regions of Luxor and Kom Ombo are not real Sudanis. They are very well paid, provided with good uniforms and rations, and a certain percentage are allowed to have their wives with them on the coast. Every camp has its 'harimat'—married quarters—where the married men and their families live. But before a man is allowed to marry he has to pass a test in musketry. Many of the men marry Arab women, and this sometimes caused considerable trouble among the Sudanese wives, who are by no means fond of their Arab sisters. As they all live close together in

**Figure 2.43** A scene from Siwa, depicted by Belgrave. *Source: Belgrave 1923.*

> rather cramped quarters they have a very lively time. One's office hours are often occupied in endeavouring to pacify some irate old Sudanese lady who brings a furious complaint that the Arab wife of her next-door neighbour is 'carrying on' with her husband.

He seemed to like his men, warts and all (p.41):

> The Sudanis of the Camel Corps are a very likeable lot. They are thoroughly sporting and have a strong sense of humour, but in many ways they are very like children. They have an aptitude for drill and soldiering but are useless without British officers owing to their lack of initiative. They are faithful and become very attached to Englishmen, but they have a keen sense of discrimination, Like all native troops there is a tendency

> for each man to consider himself a born leader, and offer his advice and opinion on all occasions; this takes a long time to subdue. But with careful training they become efficient soldiers, and they look very smart in their khaki uniform, which is rather similar to an Indian's. Physically many of them are splendid men, very powerful and muscular, like bronze statues, but although the climate of their own country is intensely hot they are by no means immune from the effects of sun, and they seem to be almost more liable to catch fever than an Englishman.

Belgrave himself, however, at one stage became seriously ill with typhoid and had to be evacuated to Mersa Matruh in an armoured car.

He describes what it was like to ride a camel (p.48):

> A good 'hagin'—riding camel—is very comfortable to ride when it is trotting, but not at a walk. Its action is peculiar, first the two off legs move together, and then the two near legs; this is what causes the swinging motion. There is an idea that when people first ride a camel they are afflicted by a sort of sea-sickness, but although I am a bad sailor I have never felt this, nor have I yet met anyone who did. Camels are very easy to ride; one just sits on the saddle with legs crossed over the front pummel, and there is very little chance of falling off as long as the camel behaves itself. The usual way of mounting is to make the camel kneel down and then step on to the saddle, but if one is long-legged and active it is possible to spring up into the saddle from the ground, which is much quicker and useful when the ground is hard and unsuitable for the camel to kneel on. A camel's usual pace is a slow trot, about 4½–5 miles an hour, and they can keep up this pace for hours on end. Of course they can go at a sort of gallop, if they like, and when they do this they cover the ground at a terrific speed, but it is rather difficult to ride them, and they have an unpleasant trick of suddenly swerving which generally shoots one over the camel's head on to the ground. I have known, too, camels that bucked, and others that suddenly knelt down when one did not expect it, both very disconcerting tricks. They are not affectionate animals and they never seem to know their own masters; there was only one among mine that had any 'parlour tricks', and he used to inhale tobacco smoke through his nose, apparently with the greatest appreciation.

After his spell in Siwa, Belgrave was an Administrative Officer in Tanganyika Territory in 1924–25. Thereafter he spent most of his life in Bahrain. He arrived there in March 1926, and remained until 1957. He was commonly referred to as *Al Mustashar*, 'the Advisor'. Always with a large cigar in hand, he advised the rulers of the island state.

## DOUGLAS NEWBOLD AND W.B.K. SHAW

Two very important people in the exploration of the southern Libyan Desert in the 1920s were Douglas Newbold and W.B.K. Shaw. Newbold (1894–1945) was born in genteel Tunbridge-Wells, the youngest of a family of eleven children, attended Uppingham (an English public school) and then went up to Oriel College, Oxford, to read Classics. That far his life was tolerably conventional. However, his Oxford career was interrupted by the First World War, in which he was commissioned in the Dorset Yeomanry and the Cavalry Machine Gun Corps. He served in Egypt and Palestine. After chasing the Senussi he took part in cavalry charges at Agagia and Mughar and was wounded at Zeitun after the fall of Gaza in November 1917. When he returned to his Oxford studies, he had already developed an interest in the Middle East and its history and archaeology. This persuaded him to join the Sudan Political Service, where he saw service in Kordofan. He arrived there in 1920 and served as an assistant district commissioner. He began to fulfil his ambition to explore the remote regions of the country, undertaking exploration by camel into the Libyan Desert in 1924. In 1927 he went across unexplored country by camel to Selima Oasis with W.B.K. Shaw. Newbold, a life-long bachelor, was protective of the interests of the people of the Sudan, who to his mind were 'some of the finest people in the world.' His archaeological, anthropological and historical interests added greatly to the knowledge of the Sudanese portions of the Libyan Desert. He became Chief Secretary of the Government of Sudan at the outbreak of the Second World War and gave assistance to the Long Range Desert Group. He was knighted in 1944 but died the following year after a riding accident in Khartoum.

As for Bill Shaw (W.B. Kennedy-Shaw, 1901–1979), after being educated at Radley and University College, Oxford, he had entered the Sudan Civil Service in 1924 as a Forestry official, but subsequently became a professional archaeologist. In 1930 and 1932 he accompanied Ralph Bagnold on motor car journeys into the Libyan Desert.

He took part in various excavations in the Middle East and from 1936 to 1940 worked for the then Palestine Department of Antiquities. He was described thus by Michael Mason (*The Paradise of Fools*, p.19):

> In person he is slight and well made; neither tall nor short; long legged for his height, active and very enduring; gentle in speech and by nature silent. His mind is tidy and academic. His natural abhorrence of showing his feelings leads him to acquit his own utterances of all passion by rendering the most human of them into Latin. Bill would pass unnoticed in a crowd until something happened; then without doing anything apparent, he would lead that crowd with efficiency.

In the Second World War he served as intelligence officer of the Long Range Desert Group, and he wrote the first history of that amazing unit. He was also with the SAS.

The first camel expedition described here was made by Newbold in 1924. This was published in that year in *Sudan Notes and Records* and he described it as 'A desert Odyssey of a thousand miles'. He went from northern Kordofan into the Dongola region of the Libyan Desert. His completed party numbered 36 men and 40 camels, including 20 Arab guides and escort, 6 police, 9 servants and hired camel men. He had 27 rifles and 6 water-camels carrying 96 gallons in iron fardasses, while the police and Arabs had their own water skins. He had with him a Kodak camera, an oil compass, a survey wheel, which was pushed by a stalwart Nuba convict from Soderi, and a good small aneroid.

After crossing the Wadi Um 'Afarit they marched over rolling pasture country, passing many herds of Kababish camels and encountering their first ostriches. He described what was termed the *gizzu* country (p.52):

> The gizzu area is simply a mass of old sand-dunes deposited, geologically speaking, in some distant 'desert' period prior to the present steppe period. When this steppe period supervened moister conditions favoured the growth of trees, bushes and grasses, whose roots and vegetable debris consolidated the surface of the shifting dunes. A later regression to the present semi-desert conditions eliminated the trees, thinned the bushes, but left enough matted vegetation to fix the general lie of the dunes and provide a yearly crop of grasses. The undulations of the dunes are gentle and their surfaces firm.

He then headed to the Wadi Hawa (Howar), an ancient river valley that used to flow from Ennedi to the Nile. He recorded archaeological sites and made impressions of rock art. They then went to Bir Natrun, where he explored the wells and salt-pans. On p.69 he described it as 'a barren desert spot with two or three large salt pans and three tiny clumps of vegetation, that can hardly be dignified by the name of oasis'. He regarded it as a 'revolting spot'.

During the winter of 1927–28, Kennedy-Shaw and Douglas Newbold, on leave from the Sudan Government service, travelled the Darb el-Arba'in slave road from El Obeid to Wadi Halfa, covering 1600 km by camel. The nature and history of this road was written up by Shaw in 1929. It derived its name – the 40 days road – from the number of days required to accomplish the journey along it, from Assiut on the Nile to El Fasher in the Sudan. The journey was written up in *Sudan Notes and Records* for 1928.

On p.104, Newbold outlined their proposed route:

> In April, 1926, Shaw and I conceived the idea of an expedition to penetrate, not only to Nukheila, but 150 miles beyond to the legendary Oasis of Zerzura, which had been reported from a number of sources for over a century as lying in the heart of the unexplored area. Hassanein's discovery of Owenat and Arkenu had shown that there were still hills and water-sources awaiting discovery, and it was quite possible that more than one oasis was concealed between Selima and Nukheila.

The journey was carried out without incidents (p.105):

> It has been said that the most successful exploration is that which has no adventures. If we had no adventures, it was partly due to good fortune, but mainly to the assistance of those friends who enabled us to start off – men, animals and equipment – in good condition. A few lame camels, a leaky fantass, an inaccurate compass, a sick man, may all precipitate disaster. But at hardly any moment of our journey did we feel any real anxiety about our safety. We also owe a debt to our Arab companions, who followed us without murmuring far out of their known range of desert, and to the camels, which showed unsuspected powers of endurance of cold, hunger and thirst. We were glad to bring them all safely to the banks of the Nile without casualty.

They started out from El Obeid on 28 October, 1927, with 15 riding and 22 baggage camels, carrying stores, water, grain, etc. They carried 17 rifles and nearly 800 rounds of ammunition. Their average speed of travel was between 2.8 and 2.9 mph. They went to the salt lake at Nukheila (also known as Merga), and then on to the small oasis of Selima, with its good quality water where camels had their first drink for 11 days. They arrived on the

Nile at Wadi Halfa on 27 December, 1927, where (p.130) 'It was a great relief to be within continuous sight of water and grass, and to be freed from the perpetual tyranny of compass and theodolite'. They had indeed carried out a great deal of position fixing, mapping and collecting, had made meteorological readings, observed evidence for formerly wetter climates, and collected archaeological geological and natural history specimens. They did not find the mythical oasis of Zerzura.

They gave detailed accounts of how their camels were managed (p.171 et seq.):

> *Watering:*
> When very heavily watered, camels are liable to damage themselves if started off immediately under big loads, and it may be worth waiting another day to let 'their stomachs settle.' Our camels did two waterless stretches of 240 miles and 275 miles.
>
> *Grazing and food:*
> Ours had grazing only to Nukheila. We carried intact to there 2,100 lbs. grain (dukhn) for 39 camels for the Nukheila-Selima stretch. On this stretch camels were grain-fed on nine nights. Average 6 lbs. per camel per night. Baggage camels given more than riding camels. Camels that are not accustomed to grain feeding are hand-fed till they learn. They should be fed off sacks, else there is waste, and grain gets mixed with sand.
>
> *Speed:*
> With the caravan walking loose, i.e., camels untied and unled, and grazing as they go, and marching six hours in the morning, three to four in the afternoon or evening, nearly 30 miles can be covered without any of the 'pressing' which is so bad for camels. The Arabs get a sleep at mid-day, and Europeans get two regular meals, 'breakfast-lunch' at 10 a.m. and dinner at 7-8 p.m.
>
> *Weather:*
> In very cold weather, such as we encountered, too early a start should not be made, as camels may founder from cold. Otherwise, of course, marching should be done in cool of early morning and evening. Bad stony going is, after heat, the' worst enemy. It slows down the caravan and tires camels enormously, even if it doesn't lame them. We carried one or two

> pairs of leather boots for footsore camels, but did not have occasion to use them. Arabs usually cut a piece of leather and sew it into the foot over the sore place, or work it into a punctured sole.
>
> *Loads:*
> Loads depend on local custom and type of camel. Some of our camels carried 480 lbs. for 600 miles. Average load was probably 350 lbs., but they lessened, of course, as water and grain were consumed.
>
> *Need for water:*
> How far can a camel caravan go without water? Ours did 275 miles, from Tamar el Gusar to Selima, ten days' hard marching, having done another waterless stretch of 240 miles, nine days' march, a fortnight previously. It is certain that our caravan would have done 300 miles (11 days) without loss of any camels, though very probably with a few stragglers and some shifting of loads.

Finally, they appended a list of food supplies that were required for the Europeans in the party. These included army biscuits, boxes of processed cheese, Quaker oats, tins of sausages, bacon and fruit, cocoa, Bovril, jam, flour, Worcester Sauce and figs. Newbold remarked (p. 178) that 'Good food means good digestion, and good digestion means optimism. Bad food means bad temper and pessimism'.

Plainly of good temper, both Newbold and Shaw were later to travel together by motor car with Bagnold's 1930 expedition.

## FRANCIS RODD

Francis Rodd (1895–1978) (**Figure 2.44**) was educated at Eton College and Balliol College, Oxford. From a distinguished family, his great-great grandfather was James Rennell, the great surveyor, who had used the regular pace of camels as a means of surveying distances (Rennell 1791). His father, Rennell Rodd, had travelled through Abyssinia by camel, going from Zeila on the Somali coast to the Ethiopian capital, to undertake a diplomatic mission to the Emperor Menelik.

During the First World War Francis served in the artillery and, since he spoke four languages fluently, as an intelligence officer. In 1917 he became a staff officer in the Middle East and served in Libya, Sinai, Egypt, Palestine and Syria.

**Figure 2.44** Francis Rodd in World War Two. *Source: Imperial War Museums TR1424.*

He finally became staff captain, Arab Bureau, at Damascus and earned a mention in dispatches. It was this war time service that motivated his subsequent career in desert travel. During his work in North Africa he developed an interest in the caravan routes of the Sahara and saw their relevance in the event of a new military conflict in North Africa arising. He also made various important contacts, including T.E. Lawrence, who became a close friend, and his fellow Balliol man, Ahmed Hassanein.

After the War Rodd joined the Foreign Office. He was sent to Rome, where his remit included Libya. He was released by the Foreign Office to go to the Sahara. He travelled twice through the central Sahara Desert (1922, 1927), and wrote about his findings in *People of the Veil* (1926). Amongst others, he encouraged Ahmed Hassanein and he travelled with Angus Buchanan. His studies of the Tuareg won him the RGS's Cuthbert Peek award in 1927 and the Founder's Medal in 1929. His obituary in the *Geographical Journal* (Anon 1978) ended with these words:

> A considerable personal shyness and reserve, one suspects, lay behind the confidence of the public man and accomplished speaker; and behind a somewhat brusque, indeed formidable, manner on occasion, a great capacity for affection, a deep sense of family ties, and an abundant capacity for staunch and generous friendship. Francis Rennell had his share of eccentricity, in ideas and in mannerisms. That tremendous flourish of a large and gaily-coloured handkerchief after snuff-taking will always be memorable. But these were part of the attractions of a man who was in some ways more of a nineteenth than a twentieth century figure.

Rodd made his name in exploration through his travels to the great Aïr Massif in the central Sahara. In 1922, accompanied by Captain Angus Buchanan (who had already been to Aïr in 1919) and the Canadian cinematographer T.A. Glover, Rodd travelled across the southern Sahara for nine months to study the Tuareg of Aïr and the Damergu region around Tanout. Although Rodd was not able to complete the Tuareg trek, as he had to resume his diplomatic duties in December 1922, Buchanan and Glover, however, did complete the journey from Kano in Northern Nigeria in 1922 to Tuugourt, Algeria, in 1923. Rodd wrote up this expedition in the *Geographical Journal* for that year.

The Aïr Mountains of Niger consist of nine almost circular massifs rising from a rocky plateau, bordered by the sand dunes and plain of the Ténéré Desert to the east. They rise to an altitude of more than 2000 m. The range hosts volcanic peaks, is incised by deep valleys, and makes up one of the world's largest ring dike structures – a circular feature born from early volcanic activity. Rodd chose to go to Aïr partly because it was an unknown volcanic area, partly because it was the home of the Tuareg, and partly because it was at the centre of some important trading routes. They left Kano in northern Nigeria and proceeded to French territory. Full details of this journey are provided by Angus Buchanan. Rodd wrote graphically of the landscapes and also of the Tuareg (*Geographical Journal*, p.88):

> Like most nomads, they are thieves and raiders by choice when an opportunity presents itself; they are poor, as only people who have no possessions can be poor, but as hard and as tough as any race there is in the world. They can travel 120 miles in twenty-four hours on one camel, and suffer the extremes of heat and cold alike, clad in a thin indigo cotton robe open at the sides, worn over the usual pair of broad Moslem trousers. When grain is available they eat it: otherwise milk and cheese of camels and goats, with the seeds of the grasses of their arid mountains, constitute their diet, only varied by an occasional dish of tomatoes and onions from their gardens, when they possess any, which is rather the exception; very rarely, sweetened tea or coffee mark red-letter days in their frugal lives. They never lose their temper, and rarely get angry, and their vices are remarkably few – none, I might almost say – for they do not smoke, and drink no fermented liquor of any sort; they rarely have more than one wife, though by Moslem practice they are allowed four. A passion for snuff, and the vanity they display in their personal appearance, are their principal failings; these, with their habit of incessantly asking for presents, excusable in a nomadic people who have so little, are the worst traits in their character, and may surely be forgiven them for the honour they pay to their women, for the love they display to their children, and for the kindness they show to their animals.

His second journey, described in the same journal in 1929, occurred in 1927 and took him among the Southern Tuareg, in Aïr and west of the plateau. It took some ten months in all. He had two companions: his brother Peter, and Augustin Courtauld (1904–1959), known best as a yachtsman and Arctic explorer, who did much of the surveying work.

They travelled extensively in Aïr doing survey work, studying the Tuareg (most of whom Rodd described as 'gentlemen'), and drawing rock paintings. Eventually they returned southwards to the Niger. Rodd was pleased with himself and his camels (p.10):

> During our journey we used our camels well and at times loaded them heavily. Yet all thirty survived. They completed the journey across the desert and were sold on the Niger as well as we could have expected in our hurry to dispose of them. Moreover, the five best we had were given away as presents to some of our men for their very excellent services. Not only did we not lose any camels by death, but we only had one with a sore back at the end of our journey, and that not a very serious case; we had started with several rather grave gangrenous sores. But we took constant care of our beasts and devoted adequate periods to pasturage and re-conditioning. I insisted on one of us three always attending the watering of our herd, not because we could do anything but follow the advice of Tekhmedin or Arab, but because when the Tuareg camel-man once sees that his owner is really keen, he takes immense trouble to do his best for the charges in which every one of his race is by nature intensely interested. And in this work nothing was ever too hard for Arab or Tekhmedin, either by day or by night, or however tired they might be after travelling or tracking strayed beasts . . . The truest thing ever said of the camel was that Providence manufactured him in a fit of abstraction. But with all that I ever find him a pleasant pastime and an efficient means of transport in his own inimitable way.

Rodd's fascination with the Tuareg is demonstrated in his 1926 book of the expeditions, *People of the Veil.* Rodd has recently been the subject of a major new biography (Boobbyer 2021).

In the Second World War he was in Military Intelligence Research (MIR) and had some input into Ralph Bagnold's activities with the Long Range Desert Group (LRDG), including the raid on Murzuk in southern Libya in 1941. Later in life he became an international banker and president of the RGS.

## THEODORE MONOD

Théodore Monod (1902–2000) was undoubtedly the greatest French desert naturalist of the twentieth century. Born and educated in Rouen, his father was a pastor of the Temple Protestant de l'Oratoire du Louvre, which Theodore also attended. He remained a devout Christian throughout his life. He also went to the Ecole Alsacienne, a Protestant foundation, and went on to take a degree in Natural Science at Paris University and, almost immediately afterwards, was appointed to a post at the Museum d'Histoire Naturelle.

His travels began in 1922 when, aged 20, he was sent by the museum on a research expedition to Mauritania. His brief was to study the fish in the region of the port of Nouadhibou. He found himself, as he put it, between two oceans – the water where he searched and the sands and the rocks of the desert which he contemplated. After a year, instead of returning to France, he extended his trip and headed south by camel to Nouakchott. Such journeys on mehari camels were termed meharées (see Monod 1937).

In 1927, he was chosen to take part in a scientific expedition across the Sahara, from Algiers to Dakar via Timbuktu. One of his best-known discoveries in what is now Mali's Kidal Region was the skeleton of the Asselar Man, dating back to Neolithic times. When on military service in a mobile camel corps at Hoggar, Algeria, in 1928–29, Monod wrote a book on the archaeology of Ahnet, where he had discovered rock drawings and pottery. Over a fifteen-month period from March 1934 to June 1935, he organized a large-scale expedition across the Western Sahara during which he visited for the first time the circular Guelb er Richat structure, also known as the 'Eye of the Sahara', in the Adrar massif of Mauritania (**Figure 2.45**). The route of this very long journey from 1934–1935 is mainly in two Saharan regions: Mauritania, from Adrar to Aouker, and northern Mali, from Timbuktu to Teghazza.

The following year he left Reggane on a camel in the direction of Taoudenni and then headed for Tindouf where he spent Christmas. In 1938, Monod went to Dakar as director of the newly-created Institut Français d'Afrique Noire (IFAN) and held this post until 1965. He made this institution (which still exists today as the Institut Fondamental d'Afrique Noire) a remarkable centre for scientific research in Africa. After the outbreak of the Second World War, in spite of his pacifism, he was ordered to spy on Italian military installations. His mission led to an expedition into the Libyan Desert, where he found more satisfaction in exploring than in observing the movements of Italian army trucks. Nevertheless, his account of the geography of northern Tibesti proved of value to the Long Range Desert Group. Between 1953 and 1964,

**Figure 2.45** Aster satellite image of the Guelb er Richat structure, which Monod visited in 1934. It has a diameter of c.40 km. *Source: NASA's Earth Observatory.*

he organized six expeditions to Majabat al-Kubra, a huge sand-covered area, half the size of France, lying between Mauritania and Mali. He wrote about the science of this area in a huge report in 1958.

A true polymath, on his many journeys he studied a vast range of topics, including meteorites and their craters, ancient lake beds, dunes (including those of the Erg Chech), stratigraphy, the travels of Alexander Gordon Laing, prehistory, rock art, pastoralism, zoology and botany, Libyan Desert Glass, and the lost oasis of Zerzura. The last of these interests caused a number of journeys in the last two decades of his life to seek the mythical Zerzura Oasis in Egypt and the Sudan (see Monod and Diemer 2000). Remarkably, from 1916 to 2000, he published 1881 volumes, summaries, articles and memoirs, including nearly 700 dedicated to the natural sciences. He accumulated over 20,000 samples during his travels. Two genera and 35 plant species, and eight genera and 130 animal species are named after him. Popular accounts of his travels are presented in *Méharées – Explorations au Vrai Sahara* (1937) and *L'Emeraude des Garamantes* (1984).

He spoke out regularly against pollution and took stands on many other subjects besides. He was fiercely outspoken on animal rights and on the question of hunting. A strict vegetarian, he once trekked 1,000 km in the Sahara without visiting a single water hole – to prove that he had sufficient stamina to do so without eating meat. He styled himself a humanist and a pacifist, declaring that he was 'violently non-violent'. He was involved in other protests, ranging from attacks on the drink and tobacco industries, bullfighting, nuclear power, child slavery, world hunger, environmental damage and poverty, fascism and French policy in Algeria. Some of his mature musings on such matters are discussed in *Terre et Ciel* (1997).

Monod loathed the petrol engine and as a consequence was particularly unhappy about the annual Paris-Dakar motor rally in which dozens of rugged vehicles roared through the Sahara, churning up the sands. He believed that the only transport compatible with desert life was the camel, travelling at three miles an hour.

For his explorations and scientific work, he was awarded the gold medals of the French Geographical Society, the RGS and the American Geographical Society. He was commander of the Legion d'Honneur, and commander of the National Order of Senegal. He dismounted a camel for the last time on 9 January, 1994. He was then 91. He made his last journey to Mauritanian's Adrar when aged 96. He died at Versailles in November 2000, aged 98.

CHAPTER THREE

# ARABIA, THE MIDDLE EAST AND IRAN

## INTRODUCTION

The Middle East has great landscape diversity. It possesses old shields of relative stability, large rift structures, the lowest point on Earth's surface, outpourings of recent volcanic lava, high mountains and escarpments, magnificent folded rocks, large domes of salt, the world's largest coastal marshes known as *sabkhas*, salty lakes, huge alluvial fans and enormous sand seas. A discussion of its geography and its explorers is provided by Hogarth (1904).

Much of this diversity and distinctiveness can be attributed to the active interaction of a number of tectonic plates. Climate also contributes to the diversity of the Middle East. The southern part is directly influenced by the Asian monsoon and may, therefore, receive summer rain and winds from the south and southeast. By contrast, the northern part comes under the influence of winter depressions that track from the west and bring winter precipitation. The north of the region tends to have greater annual precipitation totals, as do some of the higher mountain ranges in the south, such as those of Yemen and Oman. There is, however, a great tract of country in eastern Iran, southern Jordan, Syria, Iraq, and in the Arabian Peninsula, where precipitation totals are less than 100 mm per year. Extensive portions of Arabia have rainfall that is less than a third of that figure.

Lying between North Africa and Arabia is the Sinai Peninsula. It lies between the Gulf of Suez and the Suez Canal on the west and the Gulf of Aqaba and the Negev on the east, and it is bounded by the Mediterranean Sea on the north and the Red Sea to the south. The mountains in the southern part reach heights in excess of 2500 m. We know much about its geology from the survey work undertaken there by camel in the early 1920s by H.J.L. Beadnell (1927).

The Arabian Peninsula (**Figure 3.1**) itself is essentially a great tilted block, which is highest in the west and slopes gradually eastward. This pattern is broken only in the southeast by the Oman highlands. The peninsula can be divided into the following landform regions: (1) the Western and Southern Highlands; (2) the Nejd; (3) the Sand Areas (Great Nafud, Dahana, and Rub' al Khali); (4) the Syrian Desert; (5) the Persian (Arabian) Gulf lowlands; and (6) Oman.

**Figure 3.1** Arabia and neighbouring areas. *Source: Author.*

**Figure 3.2** The great dunes of the Rub' al Khali in Oman. *Source: Author.*

The Western and Southern Highlands rise in elevation from 460 to 3,700 m in Yemen. The northern segment, from the Gulf of Aqaba to south of Mecca, is known as the Hejaz. It has a general elevation between 600-900 m, with some mountains attaining 1850–2700 m. Many ephemeral streams rise in these highlands and flow onto the narrow coastal plain called the Tihama. Southward, the highlands continue into Asir but at a considerably higher elevation. The mountains of Yemen in the southwestern corner of Arabia are the highest of the peninsula. Eastward, through the Hadhramawt (Hadramaut), the rugged highlands gradually decline to 600 m in Dhofar. The Nejd tableland lies to the east of the Western Highlands and is bordered on its north, east, and south by sand seas: the Great Nafud in the north and the Rub' al Khali in the south.

Arabia's sand seas are enormous and stretch almost down the whole latitudinal extent of the peninsula from Kuwait to Oman. They account for nearly a third of its land area, covering not far short of 800,000 km$^2$. The Rub' al Khali itself (**Figure 3.2**), with an area of between 500,000 and 600,000 km$^2$, is often called the 'Empty Quarter'. The biggest active sand sea in the world, it is approximately the size of France. Some if its linear ridges can be up to 300 km long. It was almost unknown to outsiders until Bertram Thomas,

Jack Philby and Wilfred Thesiger crossed it by camel in the middle of the twentieth century.

North of the Great Nafud (Nefud) lies the Syrian Desert, which stretches into Iraq, Jordan and Syria. The most dissected part is the divide area between the Arabian/Persian Gulf and the Euphrates drainage basin.

The Arabian/Persian Gulf lowlands extend from the head of the Gulf to the fjord-like coast of the Musandam Peninsula of Oman. Largely an area of low plains (less than 180 m in elevation), the relief is broken in a few places by elevated ridges or escarpments. The immediate coast is fringed with coral reefs, bars and marshes and lagoons (sabkhas). Oman has a series of rugged mountain ranges parallel to the coast. The Jabal al Akhdar attains a height of over 3000 m.

Much of the nineteenth- and early-twentieth-century exploration of the area was undertaken by the British, and the names of notable, though sometimes forgotten, explorers include George Sadlier (Sadleir), Richard Burton, Charles Doughty, James Wellsted, Gifford Palgrave, Jack Philby, Bertram Thomas, Gerard Leachman, William Shakespear, Wilfrid Scawen Blunt and his wife Anne, Theodore Bent and his wife Mabel, Douglas Carruthers, Gertrude Bell, Freya Stark, Percy Cox and Wilfred Thesiger. Some of these individuals were undoubtedly also involved with intelligence gathering (Satia 2008).

**Figure 3.3** An 1839 lithograph of a camel caravan on the Gulf of Akaba by Louis Haghe from an original drawing by David Roberts.

The Scottish artist David Roberts toured the holy sites of the Middle East by camel in 1838–9, and camels feature in some of his landscapes (**Figure 3.3**). However, various Europeans were also active, including Carsten Niebuhr of Denmark and Germany and Johann Ludwig Burckhardt of Switzerland.

Across the Gulf lies Iran, which also is a country of great deserts, with the huge Dasht-e-Kavir (the Great Salt Desert) in the interior, the Lut desert, with its dramatic yardangs, in the south, the edge of the Seistan Basin in the east (which extends into Afghanistan and Pakistan), the rugged Makran coast, and the anticlines and synclines of the Zagros Range in the west. The surveyors of the Indo-Afghan Boundary Commission, such as G.P. Tate deserve to be better known.

## JEAN BAPTISTE TAVERNIER

Jean-Baptiste Tavernier (1605–1689) (**Figure 3.4**) was a seventeenth-century French gem merchant and traveller. A Hugeonot, he made six voyages to Persia and India between the years 1630 and 1668. At the behest of his patron, Louis XIV, he published *Les Six Voyages de Jean-Baptiste Tavernier*. This became a best seller and during his lifetime was translated into German, Dutch, Italian and English (1677). It contains much ethnographic material and detailed descriptions of places as far apart as Aleppo and Agra. He got as far east as Java and as far south as the Cape of Good Hope. By his traffic in jewels, particularly diamonds from India, and other costly commodities of small bulk, he turned his wanderings to

**Figure 3.4** Jean-Baptiste Tavernier in robes presented to him by the Shah of Persia. *Source: Tavernier 1677, Frontispiece.*

profitable account, and amassed a fortune which enabled him to purchase the Barony of Aubonne (Rothrock 1967; Burger 2017). He visited royalty, including the Shah of Persia and the Mogul rulers of India, Shah Jahan and Aurangzeb.

Inevitably, as a result of his journeys in Turkey and Persia, he had some keen observations on camels, camel masters and caravans. The caravans were composed both of camels and horses (Tavernier 1677, 46–7):

> There are two sorts of caravans. There are caravans which consist of camels, which are the most usual, in regard that camels are cheap, and for that some camels will carry as much as three horses, others as much as four or five. But among the caravans of camels, there are several horses and mules, which the merchants themselves ride upon, it being very tedious to ride upon a camel when he only goes a foot-pace, but very pleasant when he goes upon his large trot. There are other caravans that consist only of horses, and among these, if the merchant have none of his own, he may hire one. The servants ride upon those horses that are least laden, but at Smyrna you may meet with good horses very cheap, from thirty to sixty crowns. As for those persons that are either unwilling or unable to be at any expense, they make use of asses, of which there are enough to be had. Above all things, you must take care to provide pack-horses to carry your wine, for the camel-masters being Mahometans, will not permit you to lade their camels with any such liquor, that beast being particularly consecrated to Mahomet, who strictly forbad the use of wine. You put your wine in bottles made of wild goats skins, with the hairy side turned innermost, and well-pitched within.

He thought that the camel-masters were an 'insolent sort of people', who 'you shall never know how to deal with, unless you can bring them to punishment'. He arranged for one miscreant to receive a hundred Bastinadoes (foot whipping), which were to be given him upon the spot. He also described the structure of their journeys:

> The journeys of the caravans are not equal, sometimes not above fix hours travel, sometimes ten, and sometimes twelve, it being the convenience of water, which is not everywhere to be met with, that is the rule of lodging the caravan. At all times the caravan travels more by night than by day, in summer to avoid the heat, and at other times, that you may be sure to have day enough to set up your tents. For if the caravan should come to pitch in the night, it would be impossible for them to find where

> to set up their tents, to dress and look after their beasts, make ready their kitchens, and provide things necessary for so large a company. True it is, that in the depth of winter and in the great snows, they seldom set out till two or three hours after midnight, and that sometimes they stay till daybreak. But in summer, according to the journey which they intend, they set out either at midnight, or an hour after sunset. The last time I went from Smyrna the caravan consisted of six hundred camels, and almost the same number of horse. Sometimes their number is greater, so that the camels going but by one and one after another, a caravan seems to be an army, and whether it be in travelling or lodging, they take up a world of ground.
>
> The camels that go into Persia through the Northern Provinces of Turkey, travel like horses in a cart, by seven and seven, they are tied together by a cord about the bigness of a man's little finger, and a fathom long, fastened to the pack saddle of the camel that goes before, and to the head-harness of the camel that follows. Those little cords are made no stronger, to the end that if the camel before should chance to fall into any hole, the camel behind should either keep him up, or else not be pulled in after the other. And to the end that the camel-driver who leads the foremost, may know whether the other six follow him or no, the last camel has a bell about his neck, which if it cease to ring, 'tis a sign' that some one of the small cords is broken, and that a camel has got a mischance.

Tavernier was impressed by the medical value of camel milk, and by the discipline and capabilities of camels (p.49):

> The female camel bears her burden eleven months, and her milk is a sovereign remedy against the dropsy . . . So soon as the caravan comes to the place where it is to lie, all the camels that belong to one master will range themselves in a circle, and lie down upon their sore feet, so that it is but untying one cord that holds the bales, and they will slide off gently from each side of the camel. When 'tis time to lade again, the same camel comes and crouches down between the two burdens, and when they are fastened, gently rises up with the burden again, which is done without any trouble or noise.... The camel though he be of a large bulk, eats very little, contented with what he meets with upon the heaths where he more especially looks for thistles, of which he is a great lover. But it is more admirable to consider with what patience they endure drought: For one time that I crossed over the deserts, where we were

> 65 days upon the road our camels were nine days without drink. But what is more wonderful is this, that when the camel is in the heat of generation, he neither eats nor drinks for 40 days together and he is then so furious, that without great care you cannot avoid being bitten by him: And wherever he bites, he carries away a piece of the flesh. At that time issues out of their mouths a white foam, besides that there appears on each fide of their chops two bladders large and swelled, like the bladder of a boar.
>
> In the spring time all the camel's hair falls off in three days. The skin remains bare, and then the flies are very troublesome. There is no way but to tar them all over: though it be a ventrous thing to come near them at that time.
>
> A camel must be dressed as well as a horse but the camel-keeper never uses any curry-comb, but only beats him all over with a small wand, as carpets are wont to be dusted. If a camel be hurt or galled under the pack-saddle, they never apply anything to it, but only wash it well with urine, without using any other remedy.

He repeated his views on camel-drivers:

> There is as much knavery among the camel drivers, as among our Horse-Coursers. For I remember, that being once at Caspian, a Persian merchant, thinking he had bought eight good camels, was deceived in four which he thought the best: He verily believing they had been fat and in good case, but they were only blown up. For those cheats have a trick to cut a hole near the tail, of which the purchaser takes no notice, and which they know neatly how to sew up again: In this hole they will blow till they have puffed up a lean camel, that he shall appear as fat and plump as the soundest that ever fed: whereby they often deceive the quickest sights, especially when the hair is all off, and that the bare skin is all rubbed over with tar.

## JOSEPH PITTS

A little younger than Tavernier, Joseph Pitts was a remarkable, if largely forgotten individual. He was born in Exeter, probably in 1662 or 1663, and inherited a strong attachment to the sea. He was fifteen years old when he sailed to Newfoundland, but on his return he was captured by corsairs off Spain and taken to Algeria. This was in 1678. Here he was greatly abused and harshly treated.

Pitts' book of his experiences was first published in 1704, nine years after his return, with the title *A True and Faithful Account of The Religion and Manners of The Mahometans* (**Figure 3.5**). His diary was reproduced in a Hakluyt Society collection edited by Sir William Foster in 1949.

Pitts was perhaps the first Englishman to visit Mecca and to write a detailed account. Indeed, Pitts's narrative was the first and most detailed description of Islam and the manners of Muslims written by a European during the seventeenth century. Joseph Pitts was closely studied by Richard Burton, who was the next Englishman to visit Mecca. Burton's *Pilgrimage to Al-Madinah and Mecca* has a lengthy appendix (V) on Pitts, and quotes his dairy *verbatim*.

On p.359 he discusses Pitts' experience in Algiers:

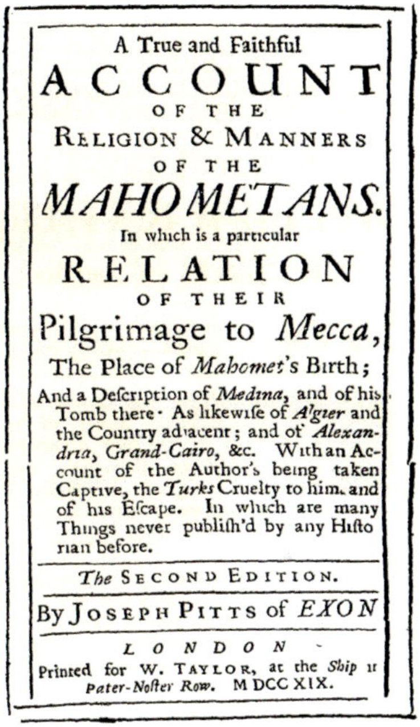

A True and Faithful
ACCOUNT
OF THE
RELIGION & MANNERS
OF THE
*MAHOMETANS.*
In which is a particular
RELATION
OF THEIR
Pilgrimage to *Mecca*,
The Place of *Mahomet*'s Birth;
And a Defcription of *Medina*, and of his Tomb there: As likewife of *Algier* and the Country adjacent; and of *Alexandria*, *Grand-Cairo*, &c. With an Account of the Author's being taken Captive, the *Turks* Cruelty to him, and of his Efcape. In which are many Things never publifh'd by any Hiftorian before.

*The* SECOND EDITION.

By JOSEPH PITTS of *EXON*

*LONDON*

Printed for W. TAYLOR, at the *Ship* in *Pater-Nofter Row*. MDCCXIX.

**Figure 3.5** Frontispiece of Pitts' book (1719 edition).

> Jos. Pitts, of Exon, youth fifteen or sixteen years old, when in A.D. 1678, his genius leading him to be a sailor and to see foreign countries, caused him to be captured by an Algerine pirate. After living in slavery for some years, he was taken by his patroon to Meccah and Al-Madinah via Alexandria, Rosetta, Cairo, and Suez. His description of these places is accurate in the main points, and though tainted with prejudice and bigotry, he is free from superstition and credulity. . . . His chief patroon, captain of a troop of horse, was a profligate and debauched man in his time, and a murderer, who determined to proselyte a Christian slave as an atonement for past impieties. He began by large offers and failed; he succeeded by dint of a great cudgel repeatedly applied to Joseph Pitts' bare feet. 'I roared out', says the relator, 'to feel the pain of his cruel strokes, but the more I cried, the more furiously he laid on, and to stop the noise of my crying, would stamp with his feet on my mouth'. 'At last', through terror, he 'turned and spake the words (la ilaha, &c.), as usual

> holding up the forefinger of the right hand'; he was then circumcised in due form. Of course, such conversion was not a sincere one – 'there was yet swines-flesh in his teeth'.

Pitts was thus converted forcibly to Islam, but he never became a Muslim at heart, and always firmly believed in the truth of Christianity (especially non-conformist Christianity). This by no means prevented him from often describing the Muslim faith reasonably accurately and even-handedly.

When his gruesome master was executed for rebellion, Joseph was acquired by a much kindlier owner, and was employed in a role similar to a house-maid. His new owner, a bachelor, decided to go on a pilgrimage to Mecca, and he took Joseph with him. They travelled via Cairo, where he found the Turks were 'much addicted to the cursed and unnatural sin of sodomy'. From Cairo they travelled by camel to Suez, and then crossed the Red Sea to Rabegh, a small port between Yambo and Jidda. They then went on to Mecca and Medina. He gives full accounts of these two holy cities. He noted the ability of the camel to live long without drinking, 'God, in his wise providence so ordering it, for otherwise it would be very difficult, if not impossible, to travel thro' the parch'd deserts of Arabia.'

Pitts (quoted in Burton, p.381) recounts his experiences with camels as he left Mecca in a caravan:

> If it happen that the camel dies by the way, the carrier is to supply us with another; and therefore, those carriers who come from Egypt to Mecca with the Caravan, bring with them several spare camels; for there is hardly a night passeth but many die upon the road, for if a camel should chance to fall, it is seldom known that it is able to rise again; and if it should, they despair of its being capable of performing the journey, or ever being useful more. It is a common thing, therefore, when a camel once falls, to take off its burden and put it on another, and then kill it; which the poorer sort of the company eat. I myself have eaten of camels flesh, and it is very sweet and nourishing. If a camel tires, they even leave him upon the place. The first day we set out from Mecca, it was without any order at all, all hurly burly; but the next day every one laboured to get forward; and in order to it, there was many time much quarrelling and fighting. But after every one had taken his place in the Caravan, they orderly and peaceably kept the same place till they came to Grand Cairo. They travel four camels in a breast, which are all tied one after the other, like as in teams. The whole body is called a Caravan, which is divided into several

> cottors, or companies, each of which hath its name, and consists, it may be, of several thousand camels; and they move one cottor after another, like distinct troops. In the head of each cottor is some great gentleman or officer, who is carried in a thing like a horse-litter, borne by two camels, one before and the other behind, which is covered all over with sear-cloth, and over that again with green broad cloth, and set forth very handsomely. If the said great person hath a wife with him, she is carried in another of the same. In the head of every cottor there goes, likewise, a sumpter camel which carries his treasures, &c. This camel hath two bells, about the bigness of our market- bells, having one on each side, the sound of which may be heard a great way off. Some other of the camels have round bells about their necks, some about their legs, like those which our carriers put about their fore-horses necks; which together with the servants (who belong to the camels, and travel on foot) singing all night, make a pleasant noise, and the journey passes away delightfully. They say this musick make the camels brisk and lively. Thus they travel, in good order every day, till they come to Grand Cairo; and were it not for this order, you may guess what confusion would be amongst such a vast multitude.

Eventually, having been granted his freedom, he managed to get back to Exeter – he had been away for somewhere between fifteen to seventeen years and was now aged just over thirty. He died around 1739.

## CARSTEN NIEBUHR

Carsten Niebuhr, (1733–1815) (**Figure 3.6**) was a German mathematician, cartographer and explorer, who came from a family of poor peasants from the marshes of Friesland. He is renowned for his participation in the ill-fated Royal Danish Arabia Expedition (1761–1767), which was mounted by King Frederick V of Denmark. Like other learned Europeans of his time, the king was aware of how little was actually known about the Arabian Peninsula. He determined to change that. Enthralled by the intellectual achievements of Prussia's Frederick the Great, Frederick eagerly responded to a proposal put to him by Johann David Michaelis, a distinguished theologian and orientialist from Göttingen University. Michaelis outlined five goals that won the king's favour. First, the expedition was to explore the geography of the Arabian

**Figure 3.6** Carsten Niebuhr in the attire of a distinguished Turkish Arabian in Yemen. *Source: Niebuhr 1792, 382.*

Peninsula; second, *en route*, it would record the plants and animals that were encountered; third, it was to describe the culture, customs and architecture of the people; fourth, it would seek to learn about diseases and remedies that were specific to the region; and finally, it was to collect as many antique books and manuscripts as possible.

The extraordinarily ambitious expedition of six men sailed in January 1761 via Marseilles and Malta to Istanbul and Alexandria. The members of the expedition then visited Cairo and Sinai before traversing the Red Sea via Jiddah to Yemen, *Arabia Felix*, which was their main destination.

In Mocha, in 1763, the expedition's philologist, Frederik Christian von Haven, died and later the same year, en route to Sana'a, its naturalist Peter Forsskål also died. In Sana'a, the remaining members of the expedition had an audience with the Imam of Yemen, but suffered from the climate and returned to Mocha. From Mocha the expedition sailed to Bombay (Mumbai). The expedition's artist Georg Wilhelm Baurenfeind died, as did the expedition's servant Lars Berggren. Both were buried at sea. The surgeon Christian Kramer died soon after landing in Bombay, leaving Niebuhr as the only surviving member. He stayed in Bombay for fourteen months and then returned home by way of Muscat, Bushire, Shiraz and Persepolis. His copies of the cuneiform inscriptions at Persepolis proved to be a key turning-point in the decipherment of cuneiform. He also visited and sketched the ruins of Babylon, Baghdad, Mosul and Aleppo. After a visit to Cyprus he made a tour through Palestine, crossed the Taurus Mountains to Bursa and reached Constantinople in February 1767. Niebuhr's maps, charts and plans constitute perhaps the greatest single contribution in the eighteenth century to the cartography of the regions he covered.

In 1774 Niebuhr published an account of his travels in *Reisebeschreibung nach Arabien und andern umliegenden Ländern*. There was an English translation, called *Travels through Arabia and Other Countries of the East* (1792). In 1964, Thorkild Hansen wrote an account of the expedition in *Arabia Felix: The Danish Expedition of 1761–1767*.

After their sojourn in Alexandria and Cairo they set off for Suez. They loaded up their camels, with amongst other things, wine *(Niebuhr 1792, 169)*:

> But we imprudently put our wine into great flasks, called in the East Damasjanes, and large enough, each of them to contain twenty ordinary bottles. These vases are very liable to be broken by the jolting of the camels, as we found by the loss of a part of our wine. It is much better to put your wine, when you are to carry it upon camels, into goat's skin bottles. This species of vessels, may, at first appear little suitable for the purpose; but they communicate no bad taste to the liquor, if the skins have been properly dressed. The same vessels answer best to carry the store of water that is requisite in travelling through dry and desart countries. My companions hired horses. But I, out of curiosity, preferred a dromedary, and found no reason to repent of my choice.

Niebuhr seems to have had a fondness of both wine and women. From Suez they visited Mount Sinai and St Catherine's Monastery. They sailed from Suez down the Red Sea to Jedda. After staying there for a while they sailed southwards to Yemen. In Yemen they undertook much of their travel on asses rather than camels, as they were often traversing rocky mountains unsuited to camels. For the first part of the journey they were treated well by the local people, but when they reached Mocha things became unpleasant. Their goods were opened, manhandled and damaged. They also fell ill. Eventually they reached Sana, stayed a while, and then returned to Mocha. From there they took a ship to Bombay. Niebuhr wrote extensively on the Hindus and the Parsees. The next major stop was at the city of Surat in southern Gujarat. He then headed to Muscat, Persia, Mosul, Syria, Cyprus and Constantinople. The portion of his journey from Mosul (now in northern Iraq) was carried out in a caravan involving 1300 camels loaded with gall-nuts from Kurdistan. To cross from Aleppo to Constantinople he joined another caravan. In this winter trek, he and the camels endured extreme cold, dealing with ice and snow. His journey finally over, he rode into Copenhagen on 20 November, 1767. His travels had been notable not only for their long duration (almost seven years) and for their hardships and loss of colleagues, but also for the vast amount of detailed, new information that he had garnered. As Kathryn Tidrick (2010, 12) put it, 'Carsten Niebuhr was the first European traveller to penetrate the interior of the Arabian peninsula and produce an articulate account of it'.

In the first decade of the nineteenth century, Ulrich Jasper Seetzen (1767–1811), a German supported by the Tsar of Russia, also visited Yemen. Disguised as a doctor, his large train of camels and his collection of mineralogical and botanical materials attracted cupidity, and in 1810 he was murdered and all his notes were lost.

## JOHANN LUDWIG BURCKHARDT

In his tragically short life, Johann Burckhardt (1784–1817) (**Figure 3.7**) was the first European to visit two of the world's greatest historical sites: Abu Simbel in Nubia and Petra in Jordan. He also made a celebrated pilgrimage to Mecca. He was born in Lausanne, Switzerland, and educated in Germany, where, at the University of Göttingen, he was taught by the great naturalist and anthropologist professor J.H. Blumenbach. However, desirous of being an explorer, he moved to England in 1806 and had an introduction from

**Figure 3.7** Burckhardt. *Source: Basel Historical Museum, Switzerland.*

Blumenbach to Sir Joseph Banks, the influential naturalist and explorer. Banks was the leader of the Association for Promoting the Discovery of the Interior Parts of Africa (commonly known as the African Association), which was founded in 1788 in London. The club was dedicated to the exploration of West Africa with the mission of discovering the origin and course of the River Niger and the location of Timbuktu. With Banks's support Burckhardt persuaded this august and aristocratic body to back a mission he wanted to make, and when they did so he prepared himself by studying languages, including Arabic, surgery and science, first in England, including Cambridge University, and then in Syria. He also learnt to behave like a Muslim, wore Arabic dress and prepared himself by taking long walks in hot weather and consuming only modest amounts of food. He called himself Shaikh Ibrahim ibn Abdullah.

In June 1812, disguised as a Syrian, he crossed Palestine to the south of the Dead Sea and entered the Nabatean city of Petra (**Figure 3.8**) through the narrow chasm for which it is so famous (Burckhardt 1822, 422–4):

> It is one of the most elegant remains of antiquity existing in Syria; its state of preservation resembles that of a building recently finished, and on a closer examination I found it to be a work of immense labour.

Later, in 1812, Burckhardt set course for Suez and Cairo. He hoped to join a caravan to Fezzan from where he hoped to explore the sources of the Niger. He never achieved this ultimate aim. Dressed in the commonest garments, as an Arab peasant or small trader, with a blue cotton blouse covering a coarse shirt, loose white trousers, a calico turban and local slippers, he travelled down the Nile Valley en route to Dongola and came across the great temples of Rameses II at Abu Simbel (Ebsambal) (Burckhardt 1819) (**Figure 3.9**).

**Figure 3.8** Al-Dayr (the Monastery) at Petra. *Source: Author.*

**Figure 3.9** Abu Simbel. Well-preserved carving of Nubian slave (left) with earring. *Source: Author.*

In 1814 he travelled by camel across the Nubian Desert to Suakin, a Sudanese port on the Red Sea. There he experienced a semoum (*simoom* or dust storm) (Burckhardt, 1819):

> I have repeatedly been exposed to the hot wind, in the Syrian and Arabian deserts, in Upper Egypt and Nubia. The hottest and most violent I ever experienced was at Suakin, yet even there I felt no particular inconvenience from it, although exposed to all its fury in the open plain. For my own part I am perfectly convinced that all the stories which travellers or the inhabitants of the towns of Egypt and Syria relate of the Semoum of the desert, are greatly exaggerated, and I never could hear of a single well authenticated instance of its having proved mortal either to man or beast. I never observed that the Semoum blows close to the ground, as commonly supposed, but always observed the whole atmosphere appear as if in a state of combustion; the dust and sand are carried high into the

> air, which assumes a reddish, or blueish, or yellowish tint, according to the nature and colour of the ground, from which the dust arises. The yellow however always, more or less, predominates . . . The most disagreeable effect of the semoum on man is, that it stops perspiration, dries up the palate, and produces great restlessness. I never saw any person lie down flat upon his face to escape its pernicious blast, as Bruce describes himself to have done in crossing this desert; but during the whirlwinds the Arabs often hide their faces with their cloaks, and kneel down near their camels to prevent the sand or dust from hurting their eyes. Camels are always much distressed, not by the heat but by the dust blowing into their large, prominent, eyes.

From there he sailed across to Jeddah, joined a caravan and made his way to Mecca. He stayed two months, suffering from a severe bout of fever, before returning to Cairo. He gave very full descriptions of Jeddah, Mecca, Medina and other parts of the Hejaz.

He described the people and drew attention to their tattoos (Burckhardt 1829). He also found that the inhabitants of the Holy City and its environs had certain vices (p.183):

> They outwardly appear more religious than their southern neighbours. They are much more rigid in the observance of their sacred rites, and public decorum is much more observed at Medina than at Mekka: the morals, however, of the inhabitants appear to be much upon the same level with those of the Mekkans; all means are adopted to cheat the hadjys. The vices which disgrace the Mekkans are also prevalent here; and their religious austerity has not been able to exclude the use of intoxicating liquors. These are prepared by the negroes, as well as date-wine, which is made by pouring water over dates, and leaving it to ferment. On the whole, I believe the Medinans to be as worthless as the Mekkans, and greater hypocrites. They, however, wish to approach nearer to the northern Turkish character; and, for that reason, abandon the few good qualities for which the Mekkans may be commended.

He described the camel caravans by which pilgrims came to Mecca (p.237):

> The Syrian caravan has always been the strongest, since the time when the Khalifes, in person, accompanied the pilgrims from Baghdad. It. sets out from Constantinople, and collects the pilgrims of Northern Asia in

> its passage through Anatolia and Syria, until it reaches Damascus, where it remains for several weeks. During the whole of the route from Constantinople to Damascus, every care is taken for the safety and convenience of the caravan; it is accompanied from town to town by the armed forces of the governors; at every station caravansaries and public fountains have been constructed by former Sultans, to accommodate it on its passage, which is attended so far with continual festivities and rejoicings. At Damascus, it is necessary to prepare for a journey of thirty days, across the Desert to Medina; and the camels which had transported it thus far, must be changed, the Anatolian camel not being able to bear the fatigues of such a journey. Almost every town in the eastern part of Syria furnishes its beasts for the purpose; and the great Bedouin Sheikhs of the frontiers of that country contract largely for camels with the government of Damascus. Their number must be supposed very great, even if the caravan be but thinly attended, when it is considered that besides those carrying water and provisions for the hadjys and soldiers, their horses, and the spare camels brought to supply such as may fail on the road, daily food for the camels themselves must be similarly transported; as well as provisions, which are deposited in castles on the Hadj route, to form a supply for the return. The Bedouins take good care that the camels shall not be overloaded, that the numbers wanted may thus be increased. In 1814, though the caravan consisted of not more than four or five thousand persons, including soldiers and servants, it had fifteen thousand camels.

He also described the Egyptian caravan (p.250):

> The Egyptian caravan, which starts from Cairo, is under the same regulations as the Syrian, but seldom equals the latter in numbers, being composed of Egyptians only, besides the military escort. Its route is more dangerous and fatiguing than that of the Syrian caravan; the road along the shore of the Red Sea leading through the territories of wild and warlike tribes of Bedouins, who frequently endeavour to cut off a part of the caravan by open force. The watering-places too are much fewer on this route than on the other; three days frequently intervening between the wells, which are, besides, seldom copious, and, with the exception of two or three, are of bad brackish water. In 1814, this caravan was composed of soldiers only with the retinue of the sacred camel, and some public officers; all the Egyptian pilgrims having preferred taking the route by

> Suez. In 1816, several grandees of Cairo joined the Hadj, one of whom had one hundred and ten camels for the transport of his baggage and retinue, and eight tents: his travelling expenses in going and coming must have amounted to ten thousand pounds. There were also about five hundred peasants, with their women, from upper and lower Egypt, who were less afraid of the fatigues and dangers of the Desert than of the Sea. I saw with them a party of public women and dancing-girls, whose tents and equipage were among the most splendid in the caravan.

In 1816, going by camel, he explored Sinai, partly to avoid an outbreak of the plague in Cairo and partly because he wanted to visit Mount Sinai and Akaba. He visited such places as Wadi Feiran, Sharm el Sheik and the Monastery at Santa Katharina. He also climbed up Mount Moses. However, back in Cairo in 1817 he contracted dysentery having eaten contaminated fish and died at the age of 32. He had a Muslim funeral and it was conducted with all proper regard to his rank as haji, scholar and sheik. His influence lived on, for among those he inspired was Gerhard Rohlfs. His diaries were published in three volumes by the African Association and are remarkable not only as a record of his travels but also of the countryside and peoples that he encountered.

Another person who visited Mecca at much the same time (1807), attached to a large camel caravan, was the Spanish explorer Ali Bey el Abbassi (1767–1818) (Bey 1816).

## JAMES RAYMOND WELLSTED

James Wellsted (1805–1842) was a British naval officer who surveyed the coastlines of the Red Sea and the interior of Oman, producing one of the first decent maps of that country (**Figure 3.10**). His two main volumes, *Travels in Arabia* (1838), are an immense store of detailed information on areas that Europeans had scarcely visited before, including the Jebel Akhdar.

On this journey, Wellsted (1838, volume 1, 6–7) (**Figure 3.11**) was impressed by the Imam of Muscat:

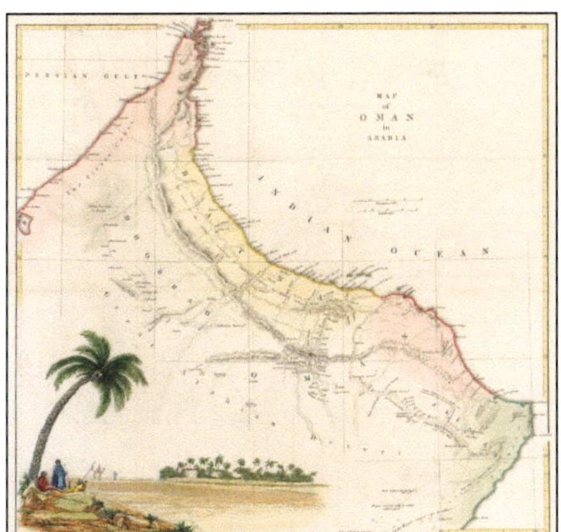


**Figure 3.10** Wellsted's map of Oman. *Source: Wellsted 1838.*

**Figure 3.11** Wellsted's route in Oman. *Source: Goudie 2016.*

> He possesses a tall and commanding figure; a mild, yet striking countenance; and an address and manner courtly, affable, and dignified. In his personal habits, the Imam has preserved the simplicity of his Bedowin origin; he is frugal almost to abstemiousness; he never wears jewels; his dress, excepting in the fineness of the materials, is not superior to that of the principal inhabitants; and he is attended, on all occasions, without pomp or ostentation. It is noticed by the Arabs, as an instance of the warmth of his affections, that he daily visits his mother, who is still alive, and pays, in all matters, implicit obedience to her wishes. In his intercourse with Europeans, he has ever displayed the warmest attention and kindness ...

Equally, he was impressed by Muscat itself (p.12):

> The town of Maskat is situated at the extremity of a small cove in the gorges of an extensive pass, which widens from this point as it advances into the interior. On either side, the cove hills, to the height of from three to five hundred feet, rise almost perpendicularly from the sea, and appear lined with forts, which, considering they belong to the vicinity of an Arab town, are in a tolerable state of repair.

Inland from the port of Sur, he received hospitality from the local women (p.47):

> I had scarcely seated myself on a skin, spread on the ground in one of these dwellings, when some young and very pretty females entered, bringing with them a huge bowl of milk. Out of compliment to them I took a long draught; but no, this was insufficient. Was it bad? – try again, and again! In vain I extolled it to the skies; I was not permitted to desist until I had swelled almost to suffocation, and sworn by the beard of the Prophet that I could and would take no more. They were then delighted, and we

> became such excellent friends that, with the assistance of a few presents and some fair speeches, we parted with expressions of mutual regret.

He also seemed to gain great exhilaration both from the Bedouin and the desert landscape (p.72):

> While sweeping across these solitary and boundless wastes, although destitute of trees, mountains, and water, or any of the features common to softer regions, there is something in their severely simple features, their nakedness and immensity, which reminds me of the trackless ocean, and impresses the soul with a feeling of sublimity. The aspect of my companion is in perfect keeping with the peculiar attributes of his native land. His sinewy form, and clean and compact limbs, are revealed by the scantiness of his garments: his dark and ruddy countenance is lighted up by the kindling of his resolute eye: his demeanour is honest and frank, and his whole appearance breathes a manly contempt of hardships.

In the mountains, he encountered the Beni Riyam people and noted the effects of alcohol (p.144):

> In their persons, although more athletic and robust than their neighbours of the plains, they have not the usual healthy and hardy look of mountaineers, but, on the contrary, their faces are wrinkled and haggard, and appear as if suffering from premature decay. I have little doubt but this is owing to an immoderate use of wine, which they distil from their grapes in large quantities, and partake of openly and freely at their several meals. They defend the practice by asserting that the cold renders it necessary.

However, one of the most interesting features of the book is his extensive discussion of camels – one of the first in the literature (p.291–303) (**Figure 3.12**). He described them as 'the most useful animal which the bounty of the Creator has bestowed on man'. He remarked that:

> Camels in all parts of Arabia are esteemed a gift of inestimable value, and those of Omán enjoy a deserved celebrity for strength and swiftness . . . the Omány in all ages is celebrated in the songs of the Arabs as the fleetest; their legs are more slender and straight, their eyes more prominent and sparkling, and their whole appearance denotes them of higher lineage than the ordinary breed of the animal.'

**Figure 3.12** An image of a camel, drawn by Joseph Bonomi and engraved by Louis Haghe. *Source: Wellsted 1838.*

He reported on the speed at which they could travel:

> Anecdotes are told of two which the Imám formerly possessed, that appear almost incredible. I have heard it asserted by those on whose testimony I should place every reliance, that one of these favourite dromedaries carried a courier from Sib to Sohar, an ordinary journey of six days, in thirty-six hours.

He was struck by their adaptations to the desert:

> All the camel's habits and instincts are adapted to its singular condition and the region which he inhabits. Let us contemplate the creature from the moment of its birth. Should this happen on a journey, the Bedowin receives it in his arms, and for a few hours places it on the mother's back; but at the first halting-place, the little stranger is put down to receive the parent's caresses, and ever after continues, unassisted, to follow her

> footsteps. Thus accustomed from its earliest age to long and toilsome journeys, little training is necessary, beyond proportioning the weight to its tender age, to inure them to the carrying of burthens; and they voluntarily kneel when about to be loaded for a journey, a position which their great height renders necessary. Kneeling is their natural state of rest; but when heavily laden on flinty or stony ground, it cannot be accomplished without pain. They then drop at once on both front knees, and in order to establish room for their hinder legs, are compelled in that condition, and whilst encumbered with the whole weight of the burthen, to plough them forward. The callosities on their joints, although nearly of a horny nature in the aged camels, seem insufficient to defend them, and it is impossible for the European to view the act without commiseration. In cause the animal to droop his neck to receive their foot, and on their raising it the rider is enabled to gain his seat, or they climb up behind: it pleases them.

He appreciated the value of their long necks:

> The great length of the camel's neck enables the animal, without stopping, to nip the thorny shrubs which everywhere abound on the Desert, and, although the spines on some are sufficiently formidable to pierce a thick shoe, the cartilaginous formation of their mouth enables them to feed without difficulty.

He also gave information on how long they could travel without being watered:

> Authorities differ as to the period the camel can endure thirst. Buffon mentions five days as an extraordinary instance; Tavernier, a good authority, nine; but it appears that camels, like several other ruminating animals, when fed on succulent herbage, do not require water; and a friend, who has had ample opportunities of judging, assures me that he once travelled from Baghdad to Damascus, a journey of twenty-five days, without the camels once drinking – a sufficiency of moisture being afforded by the abundant vegetation there found at every stage.

He assessed the size of load they could transport:

> In Egypt, when supplied with abundance of food, they carry upwards of a thousand pounds; but the ordinary burden in a caravan journey is from two hundred and fifty to five hundred pounds. The motions and paces of

> the camel are ungainly; its walk, its trot at speed, its gallop, being equally violent and disagreeable.

On the debit side, he averred:

> Notwithstanding its patience and other admirable qualities, the camel is gifted with but little sagacity, nor does it appear to be capable of forming any strong attachment to its master, although they frequently do so to one of their own kind with which they have long been accustomed to travel . . . I should say, they are the most quarrelsome brutes in existence. After the hardest day's journey, no sooner is the baggage removed than the attention of the driver is constantly required to keep them from fighting . . .

Although Wellsted's travels in Oman are his finest achievement, he also described in great detail the coastlines of Arabia and even ventured to Berbera in Somalia, and along the coast of Yemen. In 1837, suffering from severe fever in Muscat, Wellsted shot himself in the mouth with a double-barrelled gun, but survived, though horribly injured. In bad health, physically and mentally, he retired back to England and died in London in 1842, aged only 37. His achievements were recognised in his lifetime and, for example, he was elected a Fellow of the Royal Society in 1837. He wrote beautifully and is notable for the enthusiasm he displayed both for the landscapes he travelled through, the people that inhabited them, and the character, virtues and vices of the camel.

## SIR RICHARD BURTON

Sir Richard Burton (1821–1890), a gifted linguist, anthropologist, soldier, pornographer and sexologist, was one of the greatest explorers of the Victorian age. Best known for his explorations in the Great Lakes area of eastern Africa and his endeavours with Speke to find the source of the Nile, he was also a noted traveller in Arabia. Born in Hertfordshire to an Irish father and an English mother, as a boy he became accustomed to a vagabond life, was partial to duelling and swordsmanship, and 'rode, smoked, gambled and experimented with all the available forms of profligacy' (Blanch 1954, p. 21). He was sent to Trinity College, Oxford, where he read Arabic. However, he loathed both the college, his fellow students and the city, and showed his resentment by departing the college by driving his horse and coach through the college flower beds. As the wheels of his coach ploughed over them, Burton sounded a tin trumpet and blew kisses to passing shop girls.

It was determined that he should join the Indian Army, and he sailed for Bombay as an ensign in June 1842. In India he learnt many languages and became exposed to native life and picturesque landscapes. He experimented with cannabis and opium. He acted as a secret agent in Sind and mixed with the tribes, often disguised as a Pathan or Persian. One of the secret tasks with which he was entrusted by the conqueror of Sind, Sir Charles Napier, was to report on various sexual practices, including pederasty, and in due course this became the subject of a paper he wrote on the distribution of this 'Persian' vice across the 'Sotadic Belt' of the Old World. This phase in his life is beautifully evoked by Christopher Ondaatjie (1996). Tiring of army life, Burton, known to his fellow officers as 'Ruffian Dick' or the 'White Nigger', decided to visit Arabia, and in particular the holy cities of Medina and Mecca. With the support of the RGS, where Sir Roderick Murchison was an ally, and of the East India Company, he was given leave to fulfil this ambition.

In a decree of AD 629 Mohammed had ruled the holy cities to be out of bounds to all infidels, and any non-believer found there could be executed. The journey was, therefore, fraught with danger. His first aim was to be the first unconverted Englishman to go there. Some Europeans, as Burton recognised, had been there before him, such as Burckhardt, as had some announced Islamic converts. Burton also wished to establish himself as an authority on Arabia and to remove what he called 'that opprobrium of modern adventure, the huge white blot which in our maps still notes the Eastern and Central regions of Arabia'.

In preparation, Burton honed his language skills, grew a beard, shaved his head and darkened his skin with walnut juice. He also had himself circumcised in the Islamic rather than the Jewish manner. Details of this painful procedure are given in McLynn (1990, 74). He portrayed himself as an Afghan doctor and assumed the role of a dervish, recognised by Islamic culture as a role for the socially and psychologically marginal. Georg Wallin (1811–1852), the Finn, had preceded Burton to northern Arabia in 1845–1848, visiting by camel such places as Mecca and Ha'il, and also assumed the disguise of an Islamic doctor and vaccinator (Wallin 1854). Unfortuntely, he never wrote about his travels at any length, and he died very young (Trautz 1930, 1932).

In July 1853, Burton made his way from Egypt across the Red Sea to the Arabian port of Yenbu, which he found as 'in no wise remarkable', and to be inhabited by races who were 'bigoted and quarrelsome'. He travelled in a

**Figure 3.13** A camel-borne sedan litter in a caravan from Syria at Medina.
*Source: Burton 1855.*

pilgrim caravan. From there he proceeded inland on a camel with a *shugdug*, a large basket similar to a litter, normally used for carrying children, the effeminate and the elderly (Burton 1855, volume 1, 233) (**Figure 3.13**). This means of transport was necessitated by his having an infected foot from stepping on a sea urchin on the Red Sea coast. It also had the benefit of allowing him to take notes more easily than if he had been on a camel's back. He spent a month in Medina, touring the shrines, most notably the Prophet's tomb and the Prophet's Mosque (The Harim).

Burton described the buildings and foods of the city, and discusses the costs of different types of slave:

> A little black boy, perfect in all his points, and tolerably intelligent, costs about a thousand piastres; girls are dearer, and eunuchs fetch double that sum. The older the children become, the more their value diminishes . . . The Abyssinian, mostly Galla, girls, so much prized because their skins are always cool in the hottest weather, are here rare; they seldom sell for less than £20, and they often fetch £60.

At Medina he joined a 7000-strong 'Damascus Caravan' of hajis (pilgrims) and they traversed the barren and desolate terrain to Mecca, using the Darb el Sharki. This inland road had never before been travelled by a European. Burton provides a graphic description of the Damascus or Syrian caravan (p.418–9):

> Huge white Syrian dromedaries, compared with which those of Al-Hijaz appeared mere pony-camels, jingling large bells, and bearing Shugdufs (litters) like miniature green tents, swaying and tossing upon their backs; gorgeous Takht-rawan, or litters carried between camels or mules with scarlet and brass trappings; Badawin bestriding naked- backed "Daluls" (dromedaries), and clinging like apes to the hairy humps; Arnaut, Kurd, and Turkish Irregular Cavalry, fiercer looking in their mirth than Roman peasants in their rage; fainting Persian pilgrims, forcing their stubborn camels to kneel, or dismounted grumbling from jaded donkeys; Kahwajis, sherbet sellers, and ambulant tobacconists crying their goods; country-people driving flocks of sheep and goats with infinite clamour through lines of horses fiercely snorting and biting and kicking and rearing; townspeople seeking their friends; returned travellers exchanging affectionate salutes; devout Hajis jostling one another, running under the legs of camels, and tumbling over the tents' ropes in their hurry to reach the Harim; cannon roaring from the citadel; shopmen, water- carriers, and fruit vendors fighting over their bargains; boys with loud screams bullying heretics; a well-mounted party of fine old Arab Shaykhs of the Hamidah clan, firing their duck-guns upwards, or blowing the powder into the calves of those before them, brandishing their swords, leaping frantically the while, with their bright coloured rags floating in the wind, tossing their long spears tufted with ostrich feathers high in the air, reckless where they fall; servants seeking their masters, and masters their tents, with vain cries of Ya Mohammed; grandees riding mules or stalking on foot, preceded by their crowd-beaters, shouting to clear the way; here the loud shrieks of women and children, whose litters are bumping and rasping against one another; there the low moaning of some poor wretch that is seeking a shady corner to die in: add a thick dust which blurs the outlines like a London fog, with a flaming sun that draws sparkles of fire from the burnished weapons of the crowd, and the brass balls of tent and litter; and I doubt, gentle reader, that even the length, the jar, and the confusion of this description is adequate to its subject, or that any "word-painting" of mine can convey a just idea of the scene.

In Mecca, he was able to visit the Grand Mosque, the Kaabah and the sacred Black Stone, and he passed through all the rituals and ceremonials demanded of a pilgrim before returning to Egypt via Jedda. He noted than many Meccans have black concubines but that the consumption of alcohol, noted by Burckhardt, had ceased. He felt that the Meccans had some undesirable characteristics, including being covetous spendthrifts, excessively proud and coarse of language.

Burton was accompanied throughout his journey by someone he generally referred to as 'the boy Mohammed'. This youth, originally from Mecca, had met up with him in Cairo:

> He is a beardless youth, of about eighteen, chocolate-brown, with high features and a bold profile . . . His figure is short and broad, with a tendency to be obese, the result of a strong stomach and the power of sleeping at discretion. He can read a little, write his name, and is uncommonly clever at a bargain. Meccah had taught him to speak excellent Arabic . . . to be eloquent in abuse, and to be profound at Prayer and Pilgrimage.

Burton, unlike some of the other British explorers, was appreciative of the Bedouin (Badawi) of the Hejaz:

> The best character of the Badawi is a truly noble compound of determination, gentleness and generosity. Usually they are a mixture of worldly cunning and great simplicity, sensitive to touchiness, good-tempered souls, solemn and dignified withal, fond of a jest, yet of a grave turn of mind, easily managed by a laugh and a soft word, and placable after passion, though madly revengeful after injury.

His remarkable journey is written up in his *A Pilgrimage to Al-Madinah and Mecca*, first published in 1855, and which was described by R.H. Kiernan (1937, 164) as being 'marked out from all other works on Arabia by its graphic description, humour, grimness, insight into Arab and all Semitic thought, vast knowledge of Easter manners and habits, vigorous, trenchant style, and highly individualistic opinions'.

Later in life, after his near fatal debacle in Somalia (in which his young colleague Willie Stroyan was killed) and his African exploits in the source regions of the Nile, for which he received the RGS Founder's Medal in 1859, Burton had a career as a diplomat, serving is such diverse postings as Fernando Po, Santos in Brazil, Damascus and Trieste. This great man, who was striking physically and overpowering intellectually, died in that last posting in 1890, but his remarkable stone tomb, in the form of an Arab tent, lies in a Catholic cemetery in Mortlake, West London. It contains something very appropriate – camel bells.

## HENRY MARTYN FIELD

**Figure 3.14** Henry Field. *Source: Clifton Pierce 1901, 646.*

Henry Martyn Field (1822–1907) (**Figure 3.14**) was an American author and clergyman. He was born at Stockbridge, Massachusetts, and graduated from Williams College in 1838. He was pastor of a Presbyterian church in St Louis, Missouri, from 1842 to 1847, and of a Congregational church in West Springfield, Massachusetts, from 1850 to 1854. From 1854 to 1898, he was editor and proprietor of *The Evangelist*, a New York periodical devoted to the interests of the Presbyterian church.

For religious reasons, Field undertook a visit to Sinai, the results of which he wrote about in *On the Desert – Recent Events in Egypt* (1888). He went to Cairo in 1882 and then crossed the Gulf of Suez before travelling around Sinai, visiting such places as the Wadi Feiran (**Figure 3.15**), Gebel Mousa and St Catherine's monastery. He also went to Jerusalem. In some ways he was following in the footsteps of the great Scottish artist, David Roberts, who drew many landscapes of these locations during his camel tour of 1838–1839. Also in the 1830s the American traveller, John Lloyd Stephens, most famous for his work on the Mayans, visited Sinai by camel (Stephens 1839). He was also following in the steps of the Swiss Dominican friar Felix Fabri (1441–1502) who travelled from Gaza to Sinai in 1483 using camels and donkeys (Prescott 1958).

Field moved around with a small camel caravan which he had gathered together. He wrote about the selection process (p.18):

> The selection of camels is a very nice matter, as on a good beast depends much of the comfort of one's journey. Looking over the number of those lying on the ground, I picked out a young dromedary that had rather a sleek appearance. Just then Dr. Schweinfurth, who was passing in the street, came up, and gave us the benefit of his experience as an African traveller. He thought I had chosen rather a pretty creature, but advised me to spread out what she carried broadly on her back, so as to make,

> not a narrow saddle, but a space on which one could sit in Turkish fashion, with his legs under him, or change his posture at will. I observed that this was the custom of the Arabs, by which they are able to take their long marches in the desert without undue fatigue. As my camel and I were now to be on somewhat intimate relations, I approached to make her acquaintance, and even tendered her some little caressing, attempting to stroke her gently; but in an instant she swung round her long neck, and gave me a vicious snap, which warned me not to presume on any familiarities. I concluded to make no further advances, but still virtuously resolved to be a kind and indulgent master.

He then recounts the first experience he had of camel travel (p.25):

> And now, as we are fully 'at sea,' it is time to speak of the 'ship' that carries us. To-day began my first experience of camel-riding, of which I had heard such fearful descriptions, and which is to many the great terror of the desert. An English writer, the late Albert Smith, describes the sensation to be like that which one would experience in riding on a piano-stool that was mounted on the top of a hansom cab and driven over ploughed ground! Friends had told me that my back would be broken, and for the

**Figure 3.15** Wadi Feiran, Sinai – one of the locations that Field visited.
*Source: Author.*

> first hour or two I almost expected to hear the bones crack. Yet strange to say, I lived through it, and 'still live' after a month's experience of the same kind, and find camel riding not at all unpleasant. It is a long, swinging motion, and one needs to get limbered up to it. The spine must be made flexible – not a bad thing for a man who is by nature stiff-backed. Indeed I am prepared to take up the defence of the camel as a much-abused and long-suffering beast. True, I cannot boast of his looks or of his temper. He has no beauty, like the horse, with smooth, round body, arched neck, and clean limbs. The only pretty feature of a camel is his ears, which, instead of being long like a donkey's, are small like those of a mouse. But his general features are ungainly: he seems to be all back and legs. These are not graceful proportions. Nor is the absence of physical perfection compensated by his moral qualities, so that we can say 'Handsome is that handsome does'; for the camel is not an amiable beast. He is always groaning and complaining, and has a growl like a lion. But in spite of all defects of temper, he has some notable virtues. Though he has not the speed of the horse, yet when it comes to the heavy work of carrying burdens, he leaves the horse far behind. Much as camels growl when you are loading them, yet when the burden is placed upon them, though they will not skip off like a prancing steed, they will rise up and carry it all day long. In this they are like some men, who are always grumbling, but who, when it comes to the pinch, rise up under their loads and bear them manfully; while others, who are smooth and plausible and full of promises, manage to evade every irksome duty.

However, Field was less charitable about the Bedouin (**Figure 3.16**) (p.42):

> Like a child, he lives only in the present; like a child, he acts wholly upon impulse, upon the feeling of the moment; like a child, his chief delight is in telling stories and in listening to them. The tales of the 'Arabian Nights' are simply a series of brilliant pictures of what may be witnessed still, when a group gathers about a story-teller in the bazaars of Cairo or Damascus, or round any camp-fire in the desert. A people who are thus but children must be treated like children, not like full-grown men. It is useless to present to them formal propositions or arguments. I should no more think of reasoning with a Bedawee than of reasoning with a baby. Give him backsheesh, and that he can understand, but argument he cannot understand. Try to govern him by appealing to his conscience or his common sense, and you will make a dismal failure.

**Figure 3.16** Bedouin and camels together in the landscape in north eastern Sinai.
*Source: Author.*

Field did not just write about Sinai. He was the author of a series of popular books of travel. Notable were his two volumes descriptive of a trip round the world in 1875–1876, entitled *From the Lakes of Killarney to the Golden Horn* (1876) and *From Egypt to Japan* (1877). He spent the last years of his life in retirement at Stockbridge, where he died in 1907.

## LEWIS PELLY

Lieutenant General Sir Lewis Pelly (1825–1892) (**Figure 3.17**), was born near Stroud in Gloucestershire. He was educated at Rugby School and entered the East India Company service in 1840. He served in Sind and subsequently in other parts of the sub-continent. He was Secretary of Legation at the Court of Persia from 1859 to 1860, taking up the post having ridden there from Trebizond. When his tour of duty was finished, he rode back to India by way of Herat and Kandahar, alone and in British uniform, the first Englishman to be seen in the area since the Afghan War of 1838–42. He was then appointed chargé d'affaires in Teheran where he succeeded his cousin

Sir Henry Rawlinson. In May 1861, he was part of the expedition which eventually placed Bahrain under British rule. He was transferred back to Persia as Political Resident (1862–1872). In 1865 he travelled to Riyadh, and in 1868 he signed a treaty with Mohammed bin Thani which recognised the independence of Qatar. In 1873, he was appointed Agent to the Governor General of Rajputana and, *ex officio*, Chief Commissioner of Baroda, remaining in this post until 1878 when he returned to England. In 1885, by now a lieutenant general, he was elected as Conservative Member of Parliament for the North Hackney constituency. He began to use his position within the House of Commons to assist the movement in campaigning for a Women's Suffrage Bill. He died in Falmouth on 22 April 1892 and is buried in Surrey.

His journey in 1865 to Riyadh, the Wahábí capital of the highlands of Central Arabia known as Nejd, was one of his most notable exploits. His party, wearing Arab cloaks, consisted of thirty camels carrying a naval officer to undertake astronomical observations, a surgeon, an interpreter, three Indian orderlies, a Persian servant, a Portuguese cook and provisions. It was undertaken partly to fix the position of Riyadh on the map, and partly to attempt to restrain the Wahábís, whose increase of power and interference with smaller states were believed to be a political danger to Great Britain. Their chief (*imam* or *amir*) at the time of Pelly's visit was named Faisal. He greatly resented the action of the British in endeavouring to suppress the slave trade, and his feelings towards the British government and their representative were avowedly hostile. Moreover, Faisal claimed much of the entire eastern shore of the Gulf from Kuwait to Ras al-Hadd belonged to him, as he claimed it had to his ancestors. One consequence of this was that he pursued an active policy in Oman which perturbed the Sultan of Muscat, who was a long-standing ally of the government of India, who then had responsibility for the country. Then there was the French question. Napoleon III had a desire to acquire control of an Arab nation that would

**Figure 3.17** Lewis Pelly. *Source: from a portrait in* The Graphic *for 1878.*

be allied to France. At this time, too, French influence was becoming dominant in Egypt, where Napoleon's wife's cousin, Ferdinand de Lesseps, was digging the Suez Canal. Another minor motive also weighed with Pelly. He wished to gather geographical information and he quoted the president of the RGS as saying that less was known of the interior of Arabia than in the days of Ptolemy. He added, with some hauteur, in his *Report on A Journey to Riyadh in Central Arabia* (1978, 2), that:

> The Royal Geographical Society in London had recorded that no European could show himself in the Capital of Najd without probable fatal consequences: and this record was shown to me as a sort of challenge, which I was inclined to take up, since it had been my habit to consider that an English Officer, with caution and experience, can go anywhere in Asia where his duty to the Indian Government requires him.

He also wrote a paper for the *Geographical Journal* (1865). He went by camel from Kuwait to Riyadh, and then back via Hohuf and Bahrain. On p.170 he wrote:

> We moved always on camels, starting a little before daybreak, and continuing our march until sunset. The camels browsed as they went along, and on arrival at our ground were allowed to straggle in search of their scanty evening meal before being herded for the night. I may mention in passing, that it would have proved impracticable to take loaded horses at the required rate over the country to be traversed, while our camels carried ourselves and baggage a distance of between 800 and 900 miles in twenty-six marching days, with only one intervening halt of three clear days. They drank but once during the first ten days; and had I taken the precaution of leading with me a milch camel, it would have been practicable for me to deviate into wholly waterless tracts, trusting entirely to camel's milk; for it is an undoubted fact, that during the spring, when green food is comparatively abundant, both man and horse can thrive on camel's milk alone, without any other sustenance, whether solid or liquid.

After crossing tall dune country, they eventually reached their destination, where they were accommodated (p.177) in 'a garden-house set apart for Turks, Infidels, and others who smoke tobacco'.

Pelly succeeded in establishing the position of Riyadh (p.177):

> The longitude of the Wahabee capital, as determined from our housetop at Riadth by the mean of five solar observations, is long. 46° 41' 48". Circumstances rendered it inconvenient for us to take the latitude, which was eventually determined by marching 5 miles in a direction north and by east, and there halting to take the north star. This observation was further tested by an observation of the polar star taken six hours before entering Riadth, and the latitude so determined is 24° 38' 34".

He had three meetings with the ruler. In his *Report* (1978) he recorded the first of these (p.46):

> On my approaching him, the Imam rose, but with difficulty; took my hand and felt slowly all over it; and then requested me to be seated close by him on the carpet. He was quite blind, but his face was remarkable, with regular features, placid, stern, self-possessed, resigned. He looked upwards of 70 years of age, and was dressed richly but with taste, wearing over the Arab cheffiah, a turban rolled from a green cashmere shawl. His voice was well modulated, and his words calm and measured. He was dignified, almost gentle; yet you felt that he could be remorselessly cruel.

Things did not go too well (p.51):

> On the whole, I came to the conclusion that while the Imam himself was a sensible and experienced man, yet that he was surrounded by some of the most excitable, unscrupulous, dangerous, and fanatical people that one could well come across; and my position was rendered all the more delicate because the Imam himself, being totally blind, was dependent on his Secretary for all his dealings with the exterior world, and it would have been quite possible for this frivolous and vindictive mongrel, at any moment so to represent matters as to lead to grave complications. Taking into consideration, therefore, my responsibilities to Government, and to my following, together with the slowness of my means of carriage, I determined to skate no longer on such very thin ice. Accordingly I informed the Secretary callously that I should start on the following day, and had ordered my camels for the morning.

The journey onwards from Riyadh was tough, particularly through the dunes of the Nefud (Naffoods) (p.58):

> On first entering the Naffoods our camels laboured heavily, the sand being soft, the ascents steep, and the drops on the further side sometimes almost perpendicular; so that the camels had to balance themselves on the crests, and then slide bodily down. Gradually, however, the sand became harder, and scattered with brushwood. After four hours of this work we halted; the animals quite tired out. On opening the water-skins which I had procured at Riyadh, the water was found undrinkable; and this, although I had taken great care in causing the skins to be well rinsed out before starting. Whether the Riyadh people had played us some trick or whether it was owing to the ill preparation of the skins, I cannot say. In either case, we found ourselves with a waterless march of four or five days before us, and only four skins of drinkable water left.

Politically the visit had little result, for Faisal died within a year and a period of civil war ensued in Arabia. Pelly then had to visit Oman, where there was great unrest, and he decided that he had to take stern action against the people of the coastal port of Sur (Soor) (p.71) (**Figure 3.18**):

**Figure 3.18** The port of Sur in Oman, where Pelly had to take firm action against the Wahabis. *Source: Author.*

> While aiding the late Sultan it became my duty to punish the sea-port town of Soor, where the Wahabee dependents plundered our subjects and otherwise injured them, and which port has for years been the principal centre whence the piratical slave dhows sailed for the East Coast of Africa. This pirates' nest has now received a lesson which will, I trust, convince the Arabs that they cannot injure our subjects without incurring heavy chastisement. These same Soorees, reputed the bravest and most cruel of the maritime Arabs, were the main instigators of a plot to massacre me in the Zanzibar dominions . . .

Subsequently, various other British and European travellers went to Riyadh and met the Wahabis. One of these was the Dane, Barclay Raunkiaer (*Though Wahhabiland on Camelback*, edited by G. de Gaury, 1969) who went there in 1912, leaving from Basrah and returning via Hofuf and Ajer. His route was very similar to Pelly's. Another visitor was Gerard Evelyn Leachman. He was born in Hampshire in 1880 and was one of a group of British soldiers, who as tough intelligence officers, scouted around Arabia. In 1912 the RGS funded a modest expedition to central Arabia: Leachman travelled on camel from Damascus to Qasim, Riyadh, Hasa and Bahrain, a journey of nearly 2240 km. He brought back the first photographs of the Saudi royal family and developed personal links with Ibn Sa'ud himself. Alas, he was killed, shot in the back, during a revolt near Fallujah in Iraq in August 1920, aged 40. Upright, cantankerous, irritable, brusque and ascetic, he served the British Empire with courage and devotion. However, according to Lawrence he had 'an abiding contempt for everything native' and Lawrence (Canton 2011, 68) described him as a 'ruffian' and as 'a long, lean, ugly jerking man, with deliberately bad manners, a yellow, jaundiced eye, harsh-tempered, screaming and violent'.

The other important figure was the magnificently named William Henry Irvine Shakespear. Born in 1878 in Multan, in what is now Pakistan, he was in due course appointed to the political department of the government of India and served in Bandar Abbas (Persia) and then in Kuwait, where from 1909 he was political agent. It was from here that he engaged in a series of journeys into the Arabian Desert. He also became an unswerving friend of Ibn Sa'ud and was probably the first Englishman to meet him in February 1910. Armed with a sextant and a huge plate camera, a collapsible bathtub, and a supply of carefully concealed wine and whisky, his longest journey, in 1914, was by camel from Kuwait to the Suez Canal via Riyadh, a distance of over 2800 km in 111 days. About 1900 km of the journey were through unknown country. While travelling, he bathed in a special compartment of his tent, took his meals and Moselle in splendid isolation from his Arab co-travellers, and maintained his identity as a British officer. He was also an accomplished photographer. His maps and notebooks were models of accuracy and thoroughness. Fighting on the side of Ibn Sa'ud against the Rashids, he was killed in a tribal skirmish at Jarab in 1915 and his head was cut off and hung on one of the main gates of Medina (Carruthers 1922).

## WILLIAM GIFFORD PALGRAVE

Gifford Palgrave (1826–1888) was born in London. His father, originally called Francis Ephraim Cohen, was a Jew who converted to the Church of England, thereafter adopting the name Francis Palgrave. His father set up the Public Record Office, while one of Gifford's uncles, Sir William Hooker, was the founder of Kew. Giffy, as he was known to his family, was precociously bright: he knew eleven letters of the alphabet at eighteen months and was good at spelling by the time he was three. He had a glittering academic career at Charterhouse and Oxford (Trinity College, like Burton), but this did not stop him joining the army in India (like Burton). It was in India that he became a Roman Catholic and then a Jesuit missionary, first in south India and then in the Lebanon and Syria. Under the name of M. l'Abbé Michel Sohail, eventually, with Vatican permission, he became a French agent. The reason why he should want to forward the imperial ambitions of France is not clear, but the Jesuits had drawn him to the attention of Louis Napoleon; he scented missionary possibilities and felt the draw of desert air.

In 1862 and 1863, disguised as a Syrian Christian physician (he himself described himself as a 'quack'), adopting the name Selim Abu Mahmoud al-'Eis, the Quiet One, and supported financially by Napoleon III, he set off in the stifling heat of mid-summer on a journey from north west to south east across Arabia, accompanied by a Syrian Jesuit teacher called Jurayjuray (alias Barakat). Like Doughty, he felt that a medical disguise would help to allay 'native suspicion'. On the other hand, this meant he could take no scientific instruments, undertake sketching or photography, or write extensive notes. His was not a scientific expedition though he had some perceptive things to say about various natural phenomena, including dunes and water supplies. His destinations were Ha'il and Riyadh, the latter being a city that no European had ever penetrated and the heartland of Wahhabi Islam. The Wahhabis were a reformed, puritanical and fundamentalist set who regarded even orthodox Sunnis as libertarians. They were xenophobic and their fanaticism and isolation in the heart of Arabia had bred a deep distrust of outsiders. Following the succumbing of Medina, Mecca, Lhasa, Bokhara and Timbuktu to the nineteenth-century explorers, Riyadh was at the time arguably the last of the great 'forbidden cities'.

First of all Palgrave travelled by Gaza to Ma'an. Between Ma'an and Djowf, he suffered a sand storm (1865, 18):

> The horizon rapidly darkened to a deep violet hue, and seemed to draw in like a curtain on every side; while at the same time a stifling blast, as though from some enormous oven opening right on our path, blew steadily under the gloom; our camels too began, in spite of all we could do, to turn round and round and bend their knees preparing to lie down. The semoom was fairly upon us.

In Ha'il, he mixed easily with all levels of society and was able to write extensively about its character and its people. He was able to spend seven weeks in Riyadh, but to avoid imminent death had to slip away in haste. From there, he travelled through the Yemama and Hasa to Katif (Qatif), where the Arabian Gulf was sighted eight months after the departure from Gaza. He visited Bahrain, Qatar, Bundar Abbas (in Persia), parts of the Trucial States (now the United Arab Emirates), including the village of Aboo-Debee (Abu Dhabi), and Oman.

Palgrave was also not especially fond of deserts and referred to their 'hopeless, irremediable sterility and desolation' (1864, 113). He was not one of those travellers who was besotted with the Bedouin. Indeed, in contrast to town Arabs, he found them to be treacherous, ignorant, coarse, immoral, aimless and degraded. On p.3 of his book (1865) he took aim:

> . . . the favourable picture occasionally drawn by travellers of the inviolability of Bedouin good faith is not always in that exact accordance with real facts which one could desire. On the contrary, deeds of the most cold-blooded perfidy are by no means uncommon among these nomads; and strangers under their guidance and protection, nay, even their own kindred brethren of the desert, are but too often the victims of such conduct. To lead travellers astray in the wilderness till they fall exhausted by thirst and weariness, and then to plunder them and leave them to die, is no infrequent Bedouin procedure.

By contrast he remarked that the inhabitants of the oasis of Kasim (1864, 127) were a 'fine and tall race of men', whose 'braided locks falling on either side of a handsome and open countenance, give them a somewhat rakish appearance'.

He was also impressed by the admirable and tasteful crafts practised in Hasa. His most favourable comments were, however, on Oman (p.149–150):

> In no other part of Arabia did I meet with stronger marks of advanced and long-established civilization than in 'Oman. Large stone-built houses, three and even four stories high, with ample and variously carved vestibules, vaulted passages, painted walls, and copious furniture for every use; hospitality and courteous welcome, outdoing even that of Nejd; politeness in conversation, cleanliness and ornament in dress, and much else of the same nature ... Gaiety and fondness for social amusement, industry and commercial activity, are characteristic of the Omanite race: they are mild, good-humoured and cheerful. The women of this province are of remarkable and far-famed beauty ...

Palgrave was not a fan of camels. In his account of his journey (1865, 39):

> I have, while in England, heard and read more than once of the 'docile camel'. If 'docile' means stupid, well and good; in such a case the camel is the very model of docility. But if the epithet is intended to designate an animal that takes an interest in its rider so far as a beast can, that in some way understands his intentions or shares them in a subordinate fashion, that obeys from a sort of submissive or half fellow-feeling with its master, like the horse and elephant, then I say that the camel is by no means docile, very much the contrary ...

He also wrote of their capacity for revenge. He raised the possibility that on account of eating camel meat and drinking camel milk, the Bedouin had developed various points of resemblance to their camels.

What he reported about his journey in his book, was the subject of some doubt, not least in London, and in the interests of a good story he may have been guilty of some embroidery. Indeed, Philby, unjustly, thought that Palgrave never reached Riyadh and suggested that the second half of his tale at least is a gigantic concoction. On the other hand, Freeth and Winstone (1978, 152) described him as an 'Englishman of exceptional talent: an Arabic scholar, a man of cultivated and sensitive intelligence, a writer of splendidly evocative prose ...'

After his great journey, Palgrave renounced both Roman Catholicism and his links with Napoleon III, and served in the British foreign service in a range of sometimes bizarre postings, including the West Indies, Trebizond, Manila, Bulgaria, Bangkok, and Uruguay, where he died of bronchitis in 1888.

# SAMUEL BARRETT MILES

S.B. Miles (1838–1914) was a soldier, administrator and explorer, whose name is inextricably tied in with Oman. Educated at Harrow, he joined the British Indian Army in 1857, the year of the Mutiny. In November 1866 he moved to Aden with his regiment, thus beginning his association with the Middle East. The next year he was appointed Cantonment Magistrate and Assistant Resident there. He made some camel journeys into the interior of Yemen with fellow explorer Werner Munzinger (*Journal of the Royal Geographical Society of London*, 1871). Colonel Miles, as he became, applied for a transfer to the Political Service and his first appointment was to Baluchistan where he was based on the coast of Makran. In that capacity he made a camel journey along the coast from Gwadar to Karachi (Miles 1874). Then in October 1872 he was appointed Political Agent and Consul at Muscat in Oman, remaining until 1886. He was to become a recognised and sympathetic authority on Oman and its people. While holding that post he made camel treks into the country inland from Muscat, including visits to the Jebel Akhdar and Buraimi.

His first paper was published in the *Geographical Journal* in 1896. It described a journey he took starting in 1884. The gap seems to be characteristic of his writing career. Perhaps in his most remarkable journey he went from Muscat and Mattrah south to Wadi Tayyin though the Wadi Daykah gorge (the 'Devils Gap') and then east and north, returning to the coast at Quriat. Near Ghurbetat they experienced a storm, which caused a flood in the Wadi Tayyin, through which they and their camels had to wade.

Another journey, made in 1876, was to the Jebel Akhdar – the Green Mountain (Miles 1901) (**Figure 3.19**). He had seen this area from the coast (p.465), 'its dark precipitous front rising abruptly, cliff above cliff, in wild and desolate nakedness to a height of nearly 10,000 feet'. On his journey he noted the importance of springs that gushed out and provided a water supply to towns like Nakhl (**Figure 3.20**). The water is used for irrigation and is carried in a series of underground tunnels called *falaj*. Miles provided one of the first detailed descriptions of these ingenious systems (p. 488).

Yet another journey took place, much of it inland from the Jebel Akhdar, starting in 1885 (Miles 1910). He started with his party from Muttrah (p.160):

> On December 11, I proceeded by boat to Muttrah, where my followers and the camels were assembled, and there distributed the baggage for the escort into six light loads for as many camels, in order that we might travel expeditiously whenever convenient and desirable to do so.

**Figure 3.19** Mountainous terrain of Central Oman. *Source: Author.*

**Figure 3.20** An irrigation channel in the mountains of central Oman. *Source: Author.*

> The riding-camels for the party were selected with regard to their fitness to perform the whole journey, as it was not intended to change them en route, while for my own riding the Sultan kindly chose from his own stud a valuable female dromedary called Samha, which proved one of the swiftest, most easy-paced, and docile animals I ever bestrode.

He passed through such places as Nizwa, Bahla, Ibri, Jabreen, Adam, Manha, Dhank and Rustaq (Rostak). In Nizwa and Bahla he photographed and admired their splendid large forts and intricate bustling souqs with narrow alleyways surrounded by lush green lime and date groves. He also saw the towering limestone mountains of Jebel Misht and Jebel Misfah, two of the most impressive peaks in Oman.

He found that Nizwa had a wonderful array of commercial activities (p.177):

> Such glimpses of Arab life as watching the coppersmiths, braziers, dyers, and others working in their primitive Oriental manner, are always fascinating and calculated to arrest the gaze of a stranger. Among the artisans are makers of camel-saddles, potters, silversmiths, cobblers, cameleen weavers, carpenters, makers of hulwa – the national sweetmeat, for which Nezwa is famous, as it has a different flavour from that of Mascat, and is largely exported – blacksmiths, sugar or treacle makers, masons, mat weavers, and others. But the most noteworthy part of the bazaar, which is only shaded from the sun by strips of matting here and, there, and is not particularly clean, is the copper market, which though inviting by the quaintness of its wares, is repellent from the incessant noise and deafening din of the hammering going on.

He was also greatly impressed by Rustaq, once the capital of Oman (p.422):

> Rostak is a place of great salubrity, and is famous for its cool and temperate climate. Fanned as it is by the sea-breeze on one side, and sheltered from the scorching blasts of the great desert by the noble mountain which towers so majestically and precipitously on its south- west side, the air is dry and mild, and it would, like Nakhl, make an excellent sanatorium for Mascat. It owes much of its reputation, however, to its thermal springs, of which it possesses two very copious ones, ranking among the hottest in Orean. We passed them as we rode into the town, and saw that they were both enclosed in large circular tanks or reservoirs. They are much resorted to

> by the people, who have great faith in their medicinal virtues. The waters contain in solution a good deal of calcareous matter, and it is astonishing how many use them both internally and for ablution. Being close together, they probably derive their origin from the same source or fountain, and find their way through the fissures or dislocations of the rocks.

The 'calcareous matter' is illustrated in **Figure 3.21**.

On his travels, Miles was for the most part impressed by the hospitality and tolerance of the Omani people, and developed a deep appreciation of their history and culture. In many places, he was received with a formal salute in which the ordnances were fired off in greeting; the sheikhs went out to the edges of their towns to receive him and there was much dancing and shouting. The only place where he was ill-received, was Ibri (he called it Obri) (p.414):

> It enjoys the worst reputation for the lawless and predatory character of its dwellers, the tribes occupying it being one and all thievish, treacherous, and turbulent, subsisting to a large extent on the produce of their raids

**Figure 3.21** A calcium carbonate spring deposit (centre) in the mountains of Central Oman. *Source: Author.*

> and incursions into the Batinah and other districts. In a country where every able-bodied man is habitually well armed and ready to fight and plunder, the quantity of available rascaldom is pretty considerable, and the quality of ruffianism is quite in keeping.

## ANNE AND WILFRID BLUNT

Lady Anne Blunt, the 15th Baroness Wentworth (**Figure 3.22**), and her philandering, hedonistic and adulterous husband, Wilfrid Scawen Blunt (**Figure 3.23**), were great breeders of Arab horses and notable explorers of the Middle East. Like the Bents in southern Arabia and the Bakers in East Africa, they were a formidable married team. Anne (1837–1917) was a granddaughter of Lord Byron and she married Wilfrid (1840–1922), a poet and diplomat, in 1869. The relationship was not entirely successful. Anne suffered numerous miscarriages, and Wilfrid enjoyed many mistresses, often simultaneously. Peter Brent (p.147) says that his deepest and most powerful relationship was with 'Skittles' (Catherine Walters), 'one of the most famous of those *grandes horizontales* whose courtesan proficiency, displayed in Paris, laid low with passion and venereal disease so many of the kings, nobility and *nouveaux riches* of the late nineteenth century'. When he moved one of his mistresses, Dorothy Carleton, into their home in 1906, Anne decided it was time to separate from him.

It was Wilfrid's ill-health that first brought the couple to the Levant in 1873 and initiated their interest in the Arabs. Wilfrid admired the simple religious faith of the Bedouin and the lack of bureaucracy in their society. Their fascination was further whetted by an expedition they carried out from Syria to the Euphrates in 1878 (A. Blunt 1879), an expedition on which they collected some Arab horses which they took back to Wilfrid's family estate at Crabbet Park in Sussex.

**Figure 3.22** Lady Blunt with Kassida. Photographer unknown. *Public domain.*

In her book Anne was highly appreciative of her camels (p.356):

> We shed a tear or two at parting our camels, such tears as people shed who dismiss good servants on reducing their establishment. . . .The camel, whatever his faults, is certainly not ill-tempered, and his roaring is as little terrible to any but cockney ears as the lowing of a cow. Roaring is his manner of speech and need frighten no-one.

Albeit being a breeder of fine horses, she believed strongly in the advantages that camels displayed over horses in terms of their character (p. 357).

Later in the same year they set off again to Damascus with the intention of penetrating deep into Arabia. One notable feature of their journey was that they travelled without subterfuge, as Europeans openly exercising their curiosity. They carried compass and barometer and were free to take notes. They did, however, wear Arab clothes not as a disguise but with the practical intention of avoiding too much notice. They were assisted in their travel by Muhammad in Aruk, whom they had met on their Euphrates expedition, and who was desirous of finding a wife in the Nejd. They were motivated by no great geographical ambition, but by an affection for the Bedouin and the desert. On their journey to the Nejd in (A. Blunt 1881) they covered some 400 miles of stony desert between Damascus and Al-Jawf, and were ambushed in a *ghazu* or raid by Bedouin, but they were eventually able to demonstrate their friendly intentions.

They also suffered a sandstorm (chapter 4):

**Figure 3.23** Wilfrid Scawen Blunt as he appeared in *Vanity Fair* in 1885.

> We saw it looming near us in the thick air, and soon after were almost hidden from each other in the increasing darkness. The sun shone feebly at intervals through the driving sand, but it was all we could do to keep the caravan together, and not lose sight of each other. At one moment we had all to stop and turn tail to the wind, covering our eyes and heads with our cloaks, waiting till the burst was over. Nothing could have faced it. Still we were far from having any idea of danger, for there really is none in these storms, and had plenty of time to notice how very picturesque the situation was, the camels driven along at speed, all huddled together for protection, with their long necks stretched out, and heads low, tags and ropes flying, and the men's cloaks streaming in the wind, all seen through the yellow haze of sand which made them look as though walking in the air. The beasts looked gigantic yet helpless, like antediluvian creatures overwhelmed in a flood.

Thereafter they crossed the great red Nafud (Nefud) (**Figure 3.24**), which they described as having the colour of rhubarb and magnesia, to Ha'il, where they called upon the Emir, Muhammad ibn Rashid. Notorious for his cruelty, the Emir treated them with kindness.

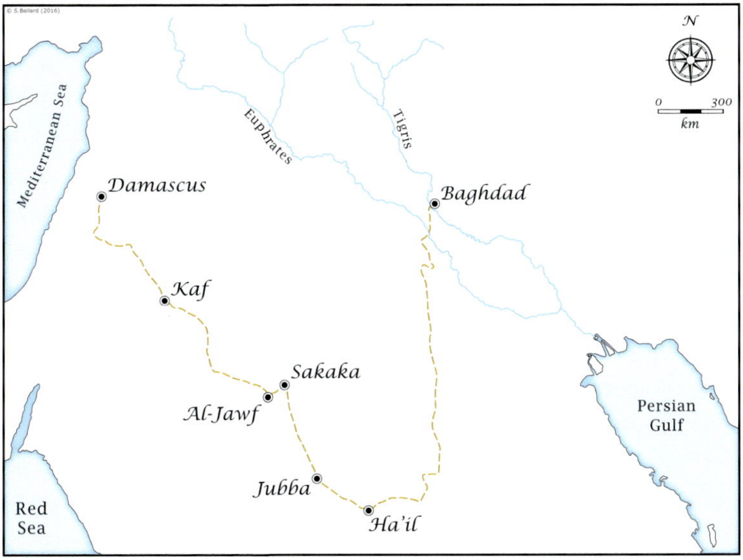


**Figure 3.24** The Blunts's route in Arabia in 1879. *Source: Goudie 2016.*

Lady Blunt's description of the Emir of Ha'il is brilliant (1881, Chapter 10):

> . . . his countenance recalled to us the portraits of Richard the Third, lean, sallow cheeks, much sunken, thin lips, with an expression of pain, except when smiling, a thin black beard, well-defined, black knitted eyebrows – eyes deep-sunk and piercing, like the eyes of a hawk, but ever turning restlessly from one of our faces to the other, and then to those beside him. It was the very type of a conscience-stricken face, or one which fear an assassin. His hands, too, were long and caw-like, and never quiet for an instant . . .

The Emir let Anne visit the harem, where she was able to meet the women. From Ha'il, they joined a huge caravan, composed of as many as 4000 camels, making for Persia, and headed for Kerbela and Baghdad, where for the first time in almost three months, they were able to sleep in a proper bed. The Blunts described the way in which Persians rode in the desert (A. Blunt 1881, 318):

> A Persian riding on a camel is the most ridiculous sight in the world. He insists on sitting astride, and seems absolutely unable to learn the ways and habits of the creature he rides; and he talks to it with his falsetto voice in a language no Arabian camel could possibly understand. The jokes cut on the Persians by the Arabs never cease from morning till night. The better class of pilgrims, and of course all the women except the very poor, travel in *mahmals* or litters – panniers, of which a camel carries two – covered over like a tradesman's van with blue or red canvas. One or two persons possess *tahteravans*, a more expensive kind of conveyance, which requires two mules or two camels, one before and one behind, to carry it. In either of these litters the traveller can squat or even lie down and sleep. The camels chosen for the mahmals are strong and even-paced; and some of these double panniers are fitted up with a certain care and elegance, and the luxuries of Persian rugs and hangings. A confidential driver leads the camel, and servants sometimes walk beside it. One of the pilgrims keeps a man to march in front with his narghileh, which he smokes through a very long tube sitting in the pannier above.

They were hugely appreciative of how the camels behaved and also noted their differences in character (p.391):

> Wilfrid's white delúl, 'Helweh' (sweetmeat), was gentle and obedient; the Meccan, 'Hamra', thoughtless and vain; 'Ghazal', affectionate, but rude and inclined to buck . . . 'Hatheran,' especially, was a camel of character. He was evidently proud of his strength and his superior understanding, and possessed a singular independence of opinion which compelled respect. It was his pride to march ahead of the rest . . .

Crossing the rivers in Persia was a problem, but they managed (p.441) and confirmed that camels could indeed swim:

> First of all the beast must be unloaded to the skin. Then a cord is tied onto the tail for one man to hold by, and another mounts his back. Thus he is driven into the water, and pushed on gradually until he loses his legs. The man on his back then floats off down stream of him, and holding with one hand by the hump, splashes water in the camel's face to keep his head straight, while the other urges him from behind. The camel seems heavier than most animals in the water, showing nothing but the tip of his nose above the surface, and he is a slow swimmer.

The Blunts were very much 'hands on' travellers. They did not travel with a large escort and a great retinue of servants. For months they had only one helper and they loaded and unloaded their camels, swam them and their horses across rivers, man-handled their baggage onto rafts made of goat-skins, carried their own guns as a precaution against bandits, pitched their own tent and shot their own suppers, and if there was nothing to shoot, caught and grilled locusts.

Anne was the first European woman to penetrate the heart of the Arabian peninsula, to cross the Nafud Desert and to visit the town of Ha'il. In 1881 the Blunts bought the estate of Shaikh Obeyd in Egypt and, until their separation, divided their time between Egypt and Crabbet Park. Anne did much to save the Arab horse, was competent in Arabic and translated some of the finest pre-Islamic verse into English. She was also a decent watercolourist who had been taught by John Ruskin, and was an accomplished violin player. She displayed bravery and perseverance, but was perhaps a little lacking in imagination. She seems not to have emulated her husband's nocturnal activity. Hogarth, the doyen of historians of Arabian exploration, says that her narrative of the journey to the Nejd is notable for sobriety and accuracy as well as for observation and sympathy. Anne died of dysentery in Cairo in 1917, aged 80, and Wilfrid followed her, aged 82, in 1922.

Wilfrid was a complex individual. A conservative English aristocrat, he had been a fierce champion of the cause of the Arabs, the Egyptians, the Irish (for which he was imprisoned) and the Indians against British domination but showed clear anti-semitic tendencies. He was, for a while, a fervent Catholic, but later in life considered converting to Islam. He is said to have ruled all those around him with the absolutism of a Turkish Sultan, and to have consumed hashish and morphine. With a body that his grandson regarded as being as beautiful as Byron's, he attracted beautiful women but also the admiration of intelligent men, such as T.E. Lawrence and George Curzon. He was the great uncle of Anthony Blunt, the Cambridge spy, one of whose co-traitors was Kim Philby, the son of another great Arabian explorer.

In visiting Ha'il, the Blunts followed in the footsteps of Charles Doughty, the Frenchman Charles Huber and the German Julius Euting. In turn, they were to be followed by Gertrude Bell.

## CHARLES MONTAGU DOUGHTY

Charles Montagu Doughty (1843–1926) (**Figure 3.25**), was born in Suffolk into a religious family. The future Arabian explorer looked so weak at birth that his father had him baptized immediately. As he grew up he was anything but weak. Originally it was intended that he would go into the Royal Navy, but he was rejected on the grounds of a slight speech impediment. A dedicated patriot, this was a disappointment to him, but as a result he attended Cambridge University and he became a geologist, writing papers on such topics as the flint implements of East Anglia and the glaciers of Norway. Doughty matriculated at Gonville and Caius College in 1861, but in 1863 he transferred to Downing College, which was more sympathetic to his geological interests and more liberal regarding attendance at lectures and chapel.

**Figure 3.25** Charles Montagu Doughty.
*Source: His* Wanderings in Arabia, *1908.*

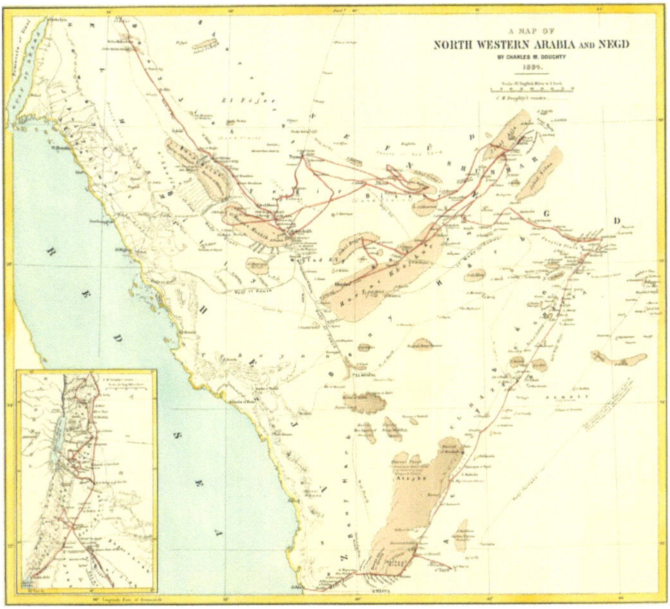


**Figure 3.26** Doughty's map of north western Arabia, showing, in red, his route, 1876–1878. *Source: Doughty 1884.*

After graduation in 1865, Doughty started to study older English literature, so that he could write an epic about early Britain as a patriotic substitute for his lost naval career. He also studied in some of the great European universities, including Louvain and Leiden, and undertook some travels in Europe and the Middle East. He also studied Arabic. In 1875 he went to Petra and learned about the Nabattean inscriptions and rock-cut tombs at Medain Salih, a place that lay deep in the desert not far from the pilgrim route to Mecca and Medina. He decided that he wanted to study these, and so in 1876 he left from Damascus to enter Arabia. He was to wander the area for almost two years, and the results of this are published in his monumental (600,000 word) *Travels in Arabia Deserta*, which appeared in 1888.

Tall, red-headed, obstinate, and with a luxuriant beard, Doughty set off from Damascus with a great pilgrim caravan (**Figure 3.26**). He carried with him no letters of introduction and was short of money. Though dressed as a Syrian and calling himself Khalil, he made no secret of being a Christian (a Nasrâny), and thus an infidel, and an Englishman. He also carried a medicine chest, reasoning that a doctor would always receive a warm welcome. First he travelled in a southerly direction to Maan (in southern modern Jordan), before reaching Medain Salih (in modern Saudi Arabia) where, fulfilling his initial aim, he copied the inscriptions that were carved in the great sandstone cliffs. However, he did not then return home. Instead, he struck out eastwards into the high desert of the Nejd to live with the nomads. In October 1877, he arrived in Ha'il, which Palgrave had visited fourteen years earlier, and the oasis of Kheibar (Khaibah), where he stayed for three and a half months. Eventually, weakened by bilharzia, he reached Taif, Jedda and the Red Sea coast in August 1878, returning to England at the end of that year.

Doughty joined his caravan, albeit late (report in *Geographical Journal,* 1884, 382–3):

> The number belonging to this caravan, according to their own computation (seldom very trustworthy, says Mr. Doughty) was about 6000 souls, together with 10,000 beasts of burden, chiefly camels, but also dromedaries, mules, and horses. The caravan guard numbered about 300 soldiers: about 100 Azeyl or dromedary riders, natives of Nejd, clad and armed as the Beduins and well fitted to deal with the nomads, who march in the rear of the caravan; while on the flanks and in the van were nearly 200 horse-soldiers, Sayal, with two field-pieces and their carriages carried upon mules.

Doughty draws a distinction, generally regarded as false, between camels and dromedaries. He reserved the latter term for riding camels and the former for baggage camels.

In *Arabia Desert* (1989 edn. 37) he wrote further about this great caravan:

> The Haj from Syria is the most considerable desert caravan of the Eastern world. There faint always some camels which have thinner soles, when these are worn to the quick in the length of so great a journey. Any such bleeding-footed beast is sold for a few crowns to the Beduins, and after some weeks' rest may be again a good camel; but if there be no buyers at hand and he must be abandoned, they cut the throat in haste, to take the skin, and go forward. The hyena, the wolf, the fox, which follow the camp, finding this meat, the carcass is rent and the bones will be scattered. I have never seen any frame of bones lying in the desert or buried by the sand-driving wind, which blows lightly and only seldom in inner Arabia.

Doughty raised the interesting issue of whether camels ever sleep (*Arabia Deserta*, 1926 edn., 221):

> The nomad Arabs say that their camels never sleep. The weary brute may stretch down his long neck upon the ground, closing awhile his great languid eyes, but after a space he will right again the great languid carcass and fall to chewing. In this fresh season they rise to graze anew in the moonlight, and roam from the booths of the slumbering nomads; but fearful by nature, they stray not then very far off.

He also described how the camels were fed (*Arabia Deserta*, 1926 edn., 65):

> The great haj camels, unlike the small Beduin cattle . . . browse nothing on the road; they are fed at the halts, as in Syria, with boiled pulse, wrought into clots of which four or five or six are crammed into the great weary beasts' jaws, and satisfy their frugal stomach.

He also referred to one bizarre use of the camel – the employment of camel urine for washing and hair care (*Arabia Deserta*, 1926 edn., 237):

> They wash their babies in camel-urine, and think thus to help them from insects: it is acrid, especially when the cattle have browsed on certain alkaline bushes, as the rimth. And in this water they all comb out their long hair, both men and women, yet sometimes thereby bleaching their locks . . .

In spite of its somewhat archaic style, which is very evident in some of the above quotes, *Arabia Deserta* became to be widely admired and is regarded as a great example of fine travel writing. It also gives a very important account of Arabian society in the last quarter of the nineteenth century, of Arabia's geology, and Nabattean archaeology. Among the great admirers of *Arabia Deserta* was T.E. Lawrence who, in an introduction to the 1921 edition, wrote:

> The more you learn of Arabia, the more you find in *Arabia Deserta*. The more you travel there the greater your respect for the insight, judgment and artistry of the author . . . It is the first and indispensable work on the Arabs of the desert . . . There is no sentiment, nothing merely picturesque, that most common failing of oriental travel books. Doughty's

> completeness is devastating. There is nothing we would take away, little we could add. He took all Arabia for his province, and has left to his successors only the poor part of specialists.

Doughty, a great patriot and Christian, who devoted his later life to family and poetry, received honorary degrees from both Oxford and Cambridge, and belatedly, in 1912, received the Founder's Medal from the RGS. He died aged 80 in Kent in 1926 and his grave is at Golders Green Crematorium. His great book came fully into its own during the First World War, when, according to his obituary in *Nature*, it was made the chief guide and work of reference for military and political staff operating in Arabia.

## THEODORE AND MABEL BENT

The Bakers and Blunts were great husband-and-wife exploring teams, as were the Bents (**Figure 3.27**). James Theodore Bent (1852–1897) was born in Yorkshire and educated at Repton School and Wadham College, Oxford, where he graduated in 1875. In 1877 he married Mabel Hall-Dare and she became his companion, photographer and diarist in all his travels. There was wealth on both sides, which provided them with the finance to travel. Mabel Virginia Anna Bent (1847–1929) came from an estate, Beauparc, on the River Boyne in County Meath, Ireland. She contributed by her skill as a photographer and in other ways to the success of her husband's journeys, and in 1900 published, jointly with her husband, *Southern Arabia, Soudan and Sakotra*, in which she recorded the results of their last expedition into those regions.

In the main, Mabel and Theodore Bent chose to spend the winter and spring months of each year travelling, using the summer and autumn to write up their findings and prepare for their next campaigns. *Inter alia*, they visited the Cyclades, Italy, Asia Minor, Ethiopia, Great Zimbabwe and Bahrain. But they were particularly noted for visiting, at considerable risk, the almost unknown Hadramut country of Yemen in 1893–1894. During this and later journeys in southern Arabia they studied the ancient history of the country, its physical features and actual condition. In the Dhofar region of Oman in 1894–1895 they visited the frankincense coast and in 1895–1896 they examined part of the African coast of the Red Sea.

While on another journey in South Arabia and Socotra (1896–1897), Theodore Bent was struck down with malaria, and died in London on 5 May 1897, just a few days after his return.

**Figure 3.27** *Left:* Mabel Bent. *Source: Bent 1892, 61. Right:* Theodore Bent. *Source: Bent and Bent 1900.*

Mabel's book, published in both their names after Theodore's premature demise, is available online and gives accounts of their journeys in southern Arabia. Many of these journeys involved travel on or with camels. The book starts with a discussion of Bahrain and especially of its famous Dilmun mounds. She then discusses visits to Muscat in Oman. Like many others they were enchanted by the city, and remarked upon its situation (p.47):

> Many-coloured volcanic rocks of fantastic form protect the horse-shoe-shaped harbour, whilst behind the white town, as far as the eye can reach, stretch deeply serrated, arid mountains, which culminate in the heights of Jebel Akhdar, or the 'Green Mountains,' some fifty miles, as the crow flies, inland, reaching an elevation of 9,000 feet.

They had two caveats (p.48):

> The shore of the town is very unpleasant, reeking with smells, and at low tide lined with all the refuse and offal of the place. At high tide shoals of fish come in to feed on this refuse, and in their train follow immense flocks of seagulls, which make the edge of the water quite white as they fly along and dive after their prey.

> The interior of Maskat is particularly gloomy: the bazaars are narrow and dirty, and roofed over with palm matting; they offer but little of interest, and if you are fond of the Arabian sweetmeat called *halwa*, it is just as well not to watch it being made there, for niggers' feet are usually employed to stir it, and the knowledge of this is apt to spoil the flavour.

Next she recounts their visit to the Hadramut. The reason for this trip may have been to gather intelligence (p. 71), for 'It was suggested to my husband that a survey of the Hadhramout by an independent traveller would be useful to the Government'. Their starting point for the interior was Makalla, which is c.380 km from Aden, and was 'the only spot between Aden and Maskat which has any pretensions to the name of port'. Mabel described the setting of the town (p.74):

> Immediately behind the town rise grim, arid mountains of a reddish hue, and the town is plastered against this rich-tinged background. By the shore, like a lighthouse, stands the white minaret of the mosque, the walls and pinnacles of which are covered with dense masses of sea-birds and pigeons; the gate of this mosque, which is really nearly in the sea, is blocked up by tanks, so that no one can enter with unwashed feet. Not far from this rises the huge palace where the sultan dwells, reminding one of a whitewashed mill; white, red, and brown are the dominant colours of the town, and in the harbour the Arab dhows, with fantastic sterns, rock to and fro in the unsteady sea, forming altogether a picturesque and unusual scene.

They then passed through Wadi Kasr and stayed for a while at Koton. Then they visited Wadi Ser, Kabr Saleh, and Shibahm, before returning to the coast at Kosseir.

Mabel discusses her travelling companions (p.78):

> For our journey inland we were entrusted by the sultan to a tribe of Bedouin and their camels. Mokaik was the name of our Mokadam or head-man, and his tribe rejoiced in the name of Khailiki. They were tiny spare men, quite beardless, with very refined, gentle faces; they might easily have been taken for women, so gentle and pretty were they. They were naturally dark, and made darker still by dirt and indigo. Their long shaggy hair was twisted up into a knot and bound by a long plaited leather string like a bootlace, which was wound round the hair and then two or three times round the head, like the fillet worn by Greek women in ancient times. They were naked save for a loin-cloth and the girdle to which were

> attached their brass powder flasks, shaped like a ram's horn, their silver cases for flint and steel, their daggers, and their thorn extractors, consisting of a picker and tweezers, fastened together.

There were also five soldiers in the party. Altogether they had 22 camels, which ate fish with some relish (p.81). They proceeded up into mountainous terrain of the Akaba (p.88):

> Words cannot express the desolate aspect of this vast table-land, Akaba or the 'going-up,' as the Arabs call it. It is perfectly level, and strewn with black lumps of basalt, looking as though a gigantic coal-scuttle had been upset. Occasionally there rises up above the plain a flat-topped mound or ridge, some 80 feet high, the last remnant of a higher level which is now disappearing. There is no sign of habitation. Only here and there are a few tanks, dug to collect the rain-water, if any falls. These are protected or indicated by a pair of walls built opposite one another, and banked up on the outer side with earth and stones, like shooting butts. The Akaba is exclusively Bedou property, and wherever a little herbage is to be found, there the nomads drive their flocks and young camels.

The next trip described was to Dhofah and the Gara Mountains. The Bents were the first Europeans to explore the mountains north of the Salalah Plain. They started from Merbat, which they found uncongenial. They gave a full description of the landscape (p.233 et seq.):

> The plain of Dhofar, along which we were now coasting, is quite an abnormal feature in this arid coast. It is the only fertile stretch between Aden and Maskat. It is formed of alluvial soil washed down from the Gara mountains; there is abundance of water very near the surface, and frequent streams make their way down to the sea, so that it is green . . .
>
> The plain is never more than nine miles wide, and at the eastern end, where the mountains were nearer to the sea, it is reduced to a very narrow strip, a grand exception to the long line of barren waste which forms the Arabian frontage to the Indian Ocean, and which gets narrower and narrower as the mountains approach the sea at Saihut. Tall cocoanut palms adorn it in clusters, and long stretches of bright green fields refresh the eye; and, at frequent intervals, we saw flourishing villages by the coast. Tobacco, cotton, Indian corn, and various species of grain grow here in great abundance, and in the gardens we find many of

> the products of India flourishing, viz. the plantain, the papya, mulberries, melons, chillis, brinjols, and fruits and vegetables of various descriptions ...
>
> The Gara mountains are now one of the wildest spots in wild Arabia; owing to the disastrous blood feuds amongst the tribe and the insecurity of travel, they had never previously been penetrated by Europeans: all that was known of the district was the actual coast-line. Exciting rumours had reached the ears of Colonel Miles, a former political agent at Maskat, concerning lakes and streams, and fertility unwonted for Arabia, which existed in these mountains, and our appetites were consequently whetted for their discovery.
>
> In ancient times this was one of the chief sources of the time-honoured frankincense trade, which still maintains itself here even more than in the Hadhramout.

On p.245 she describes the camels of the region and their gastronomic habits:

> Camels in Dhofar are not very choice feeders, and have a predilection for bones, and if they saw a bone near the path they would make for it with an eager rush extremely disconcerting to the rider. Fish, too, is dried for them and given them as food (called *kei* by the Gara and *ohma* by the Arabs), as also is a cactus which grows in the mountains, which is cut into sections for them. They are fine sturdy animals, and can go up and down hill better than any camels I have ever seen. The fertile Gara range is a great breeding place for camels, but as there is no commerce or communication with the interior, the Bedouin do not make much use of them themselves, but sell them to their neighbours, who come here to purchase.

The Dhofah experience was also written up by Theodore in the *Geographical Journal* (1895).

After describing their sojourn on the Red Sea coast of Sudan, Mabel turns to their final expedition – to the island of Socotra. They toured all over it, much of it with camels. She described it as being (p.343):

> Cast away in the Indian Ocean, like a fragment rejected in the construction of Africa, very mountainous and fertile, yet practically harbourless, the island of Sokotra is, perhaps, as little known as any inhabited island on the globe.

She noted the island's Bedouin (p.366):

> The Bedou is decidedly a handsome individual, lithe of limb like his goats, and with a *café-au-lait*-coloured skin; he has a sharp profile, excellent teeth; he often wears a stubbly black beard and has beautifully pencilled eyebrows, and, though differing entirely in language, in physique and type he closely resembles the Bedouin found in the Mahri and Gara mountains. Furthermore, the mode of life is the same – dwelling in caves when necessary, but having permanent abodes on the lower lands; and they have several other striking points in common.

They visited the highest point of the island (p.378):

> Sokotra without Mount Haghier would be like a body without a soul. The great mass of mountains which occupies the centre of the island rises in many jagged and stupendous peaks to the height of nearly 5000 feet. At all seasons of the year it catches the fugitive sea mists which so rarely visit the Arabian coasts, and down its sides flow sparkling streams and bubbling cascades. The Ghebel Bit Molek (a name which, by the way, sounds as if it had an Assyrian origin) is the highest peak. It is very sheer and unapproachable at its summit, and though only 4,900 feet high will give trouble to the adventurous crag- climber who is bent on conquering it. Then there are the Driat peaks, the Adouna peaks, and many others piercing the sky like needles, around which wild goats and civet cats roam, but no other big game.

As usual they collected vegetation and made archaeological observations. From Socotra they returned to south west Yemen, to the country of the Fadhli. It was here they contracted malaria. They struggled on camel back to the coast. She concluded her book poignantly thus (p.429):

> Our last moments at Shukra were spent lying on the sand with our heads on a bag, and sheltered by a little bit of sacking on three sticks. The sultan sat over us on a high chair, saying very polite things. We were lifted on board our ship at three o'clock, and from the ship admired Shukra, which looked very picturesque in the evening haze, with its towers, its few trees, and its many-peaked Goddam mountains behind. We reached Aden at three next afternoon. This is all I can write about this journey. It would have been better told, but that I only am left to tell it.

From Aden they returned to England, where Theodore died, leaving Mabel to write up their remarkable travels.

## JOHN FRYER KEANE

John 'Jack' Fryer Thomas Keane (1854–1937), born in the Yorkshire port of Whitby, was a clergyman's son who went to sea at the age of twelve. At the age of fourteen, he was living with wealthy relatives in Madras, where his father had been a missionary. Thereafter he made many sea voyages, to the South Seas, the Arctic, British Guyana, Bengal and the Cape of Good Hope. In late summer 1877, still only 23, he travelled to Alexandria by steamer, and from there went to Jeddah, then visiting Mecca and Medina, one of the few Europeans ever to have done so at that time. In Jeddah he joined a party who were making a pilgrimage similar to that shown in **Figure 3.28**. He passed himself off as a pilgrim. He appears to have mastered Islamic ritual and prayer, partly by his observations on his sea voyages and partly by copying what he saw.

**Figure 3.28** Pilgrim camp at Mount 'Arafah, near Mecca, in 1889, just after Jack's visit. The photo was taken by C. Snouk-Hurgronje. *Source: Snouk-Hurgronje 1889.*

Jack set off from the port of Jeddah on the Red Sea. He described his preparations (Keane 1881a, 4):

> After a great deal of hard swearing in a Babel of tongues, the camels were laden and got under way, and I, following the example of my companions, performed the greater ablution (i.e. the bathing of the entire person), and put on the pilgrim 's garb, which consists of two pieces of white or light-coloured cotton fabric, one piece round the body over the left shoulder, leaving the right arm bare, the other round the loins, enveloping the body from the waist to below the knee (the head is left uncovered and the upper part of the foot bare), and then said a two-prostration prayer, which is a supplementary act of devotion for special occasions. We afterwards joined our camels, which had become part of a larger caravan outside the walls of Jeddah. I had no difficulty in getting through the gate in the crowd, nor did I hear of any pass being required from the pilgrims, either here or at Meccah, though one may have been got for me by the Amér as Abdur Mohammed, the name I had adopted.

He described the *shugduf* in some detail (p.6):

> The kind of panniers or litters, in which we sat or lay at full length when riding, are known by the names of shugduf or shibriyah. Inside the shugduf is a pocket for food, tobacco, and other little necessaries; outside, at the tail end and within easy reach, a wicker-basket containing a bottle of water. Having mounted, that is to say, having, by the help of a ladder, ascended into this construction of bent boughs and old bags, resembling a rickety wigwam poised airily on the hump of a camel, the impression I laboured under was that we were instantly to plunge off in mad bounds across an exceedingly rough country, feeling at the same time pity for the poor beast under what I supposed could only have been an immense burden, spiced with momentary expectation of a spill.

Keane, although a sailor, found the motion of the camels was nauseous (p.7):

> I, for my part, spent one of the most wretched and apparently perilous nights I have ever had the misfortune to undergo, for the pitching and rolling of that desert-ship could not find its equal afloat; so that I, old sailor as I was, became exceedingly sick, a thing which had not happened to me for many years while at sea.

Keane adopted a disguise (p.16/17):

> The costume I assumed was a sort of mixture of Hindi and Turk, with perhaps as much of the Arab as of either in it. My headdress was a Turkish tarbouche, with a long silk handkerchief tied round it; on my feet the Hindi sandals, a strip of leather protecting the entire sole of the foot, and kept on by a band over the instep and down between the big and second toes; by way of breeches, a pair of Hindi cotton pyjamas, four feet round the waist and tight at the ankles; a tunic of the same material, girded about the loins with several folds of a cotton scarf; and over all, when in the open air, an Arab-brown cloth mantle, having wide sleeves and reaching down to the ankles.

After his sojourn in Mecca, Jack set out for the other great holy city – Medina (*My Journey to Medinah*, 1881b). He described the nature of the tow ropes and halters that were used to keep the camels in train (p.15):

> The tow-ropes are made of small cocoanut-fibre line, about an inch in circumference, and not so strong but that it will part if the camel falls; though with a terrible jar on the animal. The halters of such camels as are inclined to lag are fitted with an ingenious spurring- curb, formed of two small plates of iron with serrated edges, and these edges are turned in, and their sharp points pressed against each side of the lower jaw by the tightening of the halter.

On p.16 he discussed the mounting of camels and once again outlined the nature of the *shugdug* ('shugduf') (which earlier had been used by Richard Burton):

> But in getting into the large cumbersome 'shugduf', or covered-in litter used by pilgrims, the platforms of which are on a level with the top of the hump, great care must be taken to land in the centre, or you may overbalance and bring down the whole concern; and after getting in you must wait until your partner is mounted, so that you may deposit yourselves carefully and simultaneously on your respective sides. Then, after a little trimming of ballast from one side to the other, you may stretch out at full length, with your feet to the open or front end of the shugduf. Both my companion and I managed this without much difficulty, considering it was our first attempt of the kind, and then remarked that we were very

> hungry. This was a bad look-out; we had neither of us eaten anything since morning, and there was no knowing when we should get our next meal. Soon we both noticed a most disagreeable smell, and after some conjecture concluded it must come from the camel.

On the journey they suffered a fatal ambush (p.37):

> The news soon came up from the rear, passed from camel to camel, that the two last camels on the line had been swept off in a sudden rush of about a dozen men from an ambush, and that they had driven the camels rapidly off into the darkness, one of the riders, a man, having been able to throw himself out of the shugduf, and escape, leaving two women and a young lad in the hands of the robbers. It appeared that nothing could be done to help them, and shortly after the caravan was put in motion. The affair gave me and my companion an animating new topic of conversation. If the man had not escaped, the third warrior, who believed the whole story implicitly, would have found me more difficult to convince. As it was, I scarcely believed the whole thing. I could not bring myself to realise that two women and a boy were being murdered in cold blood within a mile or two of me.

They also suffered from camels falling down on steep slopes and breaking their *shugdugs* (p. 51), from floods (p.176) and from sand storms (p.122):

> The wind increased steadily, and the Bedawi exerted themselves more and more to push forward the caravan. But when I found that at the soonest we could not reach the end of the march under four hours I felt we were in for it, whatever it was. It proved to be what an old sailor would call a 'buster'. The view soon became obscured by a yellow haze of sand, and the sun came down close over our heads in the form of a glaring dull-red ball, in that way so familiar to a Londoner during the fogs. The wind yelled over the high rugged rocks, swept them clean, and pounced down on us in the valleys, with its burden of dust and sand, rattling a shower against us that sounded on the outside of the shugduf like hail. The poor pedestrians must have suffered tortures from the sand blowing against their bare legs and faces, for it pricked like needles, as I found by holding my hand out of the shugduf. They sheltered themselves as much as they could by getting on the lee-side of the camels and covering up their faces. The Bedawi did not seem to mind it so much. Their dress can

> be so arranged as to cover all but the lower parts of their legs, and their calves and shins seemed to be hardened to it. The air soon became so thickly laden with detritus that you could not see more than four or five camels ahead. The fine dust searched into the shugduf, and lodged in every wrinkle and fold in little ridges all over our clothes. The camels, the shugdufs, the men walking, the few green shrubs by the way, all became of the same light-yellow hue, spangled with gold. Though the wind was blowing with great violence the heat was intense.

He decided that he did not like camels (p.105):

> The camel has no hump. Well, not exactly that; but the hard-worked, ill-conditioned camel of the Hejaz has very little hump, a mere slight elevation of the dorsal ridge, more thickly clothed with wool than the rest of the body, over which the saddle fits. The saddle is a long bag of coarse sacking about the thickness of a large pillow and twice the length of the camel's back; when put on it is doubled round above the tail and the ends kept together at the shoulders by a wooden fixture constructed much on the same principle as our pack-saddles. This huge ungainly brute is endowed with a stupidity and helplessness perfectly pitiable, and its temper is diabolical. While being laden it gives vent to spuming, spluttering, bellows and whines not unlike what the roars of an enraged croupous lion might be supposed to be. I never recall to my mind a camel, but I picture it biting, or trying to bite something, an operation for which its peculiar dental arrangement well adapts it. It never loses an opportunity of biting anything it dares bite, and it dares bite anything but a Bedawi: some will even occasionally venture a sly nibble at their masters, by whom they are brutally treated. They are seldom to be seen without saddle sores, raw ulcers as often covering a surface of feet as inches, and they always die in harness. Though the camel if allowed will drink freely twice a day, I have myself ridden them three days from water to water, during which their great sufferings were most evidently exhibited. In some places the road, over an otherwise trackless plain, was literally marked out by lines of their bleached skeletons; many hundreds of which I have seen, and have often pulled the huge bones apart.

From Medina, Jack returned via Mecca to Jeddah, and thence back home. Once in England, he published two books about his visit to the Hijaz: *Six Months in Meccah* and *My Journey to Medinah*. They appeared in 1881. From

1879 to 1883 he was a journalist in London, was sent to Indo-China and he visited America in 1887 and 1890. By 1890 he had arrived in Australia. He was a member of various expeditions into the interior. He died in 1837 at Charters Towers, Queensland, in his eighty-third year. His obituary in *The Times* (Anon 1837) finished by saying that 'He was adventurous to the point of recklessness, and being indifferent to his own interests he never sought any recognition'.

## PERCY COX

Sir Percy Cox (1864–1937), an Essex boy, was educated at Harrow and Sandhurst. He became a great soldier – reaching the rank of Major-General – and a hugely influential colonial administrator. In October 1899 the new Viceroy of India, Lord Curzon (himself no mean explorer and an expert on Persia), appointed Cox as Political Agent of the government of India and consul at Muscat, Oman – a post previously held by Samuel Miles. He remained there until 1904. It was there that he undertook some important expeditions. Later in life he was a major figure in the establishment of Iraq, and he was a friend to other explorers including William Shakespear and Gertrude Bell. After Iraq, Cox devoted much of the rest of his life to the RGS, serving as its president from 1933 to 1936. In 1937, he died suddenly from heart failure while out hunting in Bedfordshire. His father had suffered the same fate. Wilfred Thesiger, who read a paper to the RGS in 1934 when Cox was president of the Society, described him as an 'awe-inspiring man renowned for his ability to keep silent in a dozen languages'. His obituary in the *Geographical Journal* (Anon 1937, p.1) said that:

> Naturally modest and self-effacing, his reputation in the councils of the great Badawi tribes of Arabia, of the shaikhs of the Persian Gulf coast, and of the political leaders in Iraq was greater than that ever reached by any other British officer. It was earned unostentatiously by the inspiration of high example and a sympathy that enabled him to see the other man's point of view, combined with a strong personality, a dignified bearing, and an honesty of purpose that was never doubted.

In May 1902, Cox made an important overland journey from Abu Dhabi on the Trucial Coast to Muscat, and was of the first Europeans to travel in the interior of Oman. In 1905 he made a similar journey, but starting from Ras

Figure 3.29 Sand dunes inland from Ras al Khaimah. *Source: Author.*

Figure 3.30 A portion of Cox's map, showing his routes (solid black lines) to Muscat from Abu Dhabi and Ras al Khaimah. *Source: Cox 1925.*

al Khaimah and back (**Figure 3.29**). During these journeys (**Figure 3.30**) he climbed the Jebel Akhdar, Oman's highest mountain. Cox was also one of the first Europeans to explore the remote and mountainous Musandam peninsula (**Figure 3.31**). He did this by *dhow* as much of Musandam was, and still is, only accessible from the sea. Another camel user at this time was the American missionary, Samuel Marinus Zwemer (1867–1952), who crossed northern Oman from Abu Dhabi and Sharja to Sohar on the Batina coast of the Gulf of Oman (Zwemer 1902).

Cox only wrote up his inland travels in the *Geographical Journal* in 1925. He cannot be accused of rushing into print.

He was enchanted by Muscat (p.194):

> The harbour and approaches to Muscat from the sea form an extra- ordinarily attractive picture, and for me possess a charm entirely their own. The first indication a traveller gets of the place as the steamer comes up the coast from the southward is a glimpse of the old Portuguese fort at the south-eastern corner of the harbour, now known as Fort Jalali. Then passing on towards the eastern arm of the harbour, or cove, as it really is, and rounding the island promontory you see on your port bow the barren rocks bearing the time- honoured record of ships which visited the port of old and painted their names on the barren rocks. Then you have before you the front and entrance of the same old fort, Jalali, with its battery of antiquated cannon directed from their embrasures at random across the harbour, while the closely packed town lying at the toe of the horseshoe opens out to view as the ship passes in between the points.

As he and the Sultan of Oman sailed round to Abu Dhabi, they passed Mussandam (p.198):

**Figure 3.31** One of the Musandam inlets near Elphinstone. *Source: Author.*

> Near the end of this promontory are two deep bays, one on either side of it. Locally, of course, they have their own Arabic names, but on our charts, and among Englishmen, they are known as the 'Malcolm' and the 'Elphinstone' inlet respectively. They are very much alike the one to the other, but the 'Elphinstone' Inlet, which wanders inland for nearly 10 miles, is of the more interest to us, as it contains the ruin of a British cable station which was located there in the first half of the nineteenth century. The scenery in the inlet reminds one greatly of a Norwegian fjord, and we may explore it in passing. Far inside is to be seen the long-deserted telegraph building. The Indo-European Telegraph Department to which it belonged found they could not keep their employes alive, much less in health, owing to the intolerable heat of the summer aggravated by the solitary conditions of life, and the station accordingly had to be transferred to the island of Hanjam, where we now have an important cable and wireless station.

He regarded Abu Dhabi as a benighted spot (p. 200). It was here he collected his camels in 1902. Early on (p.202), an unfortunate incident occurred:

> At nightfall we bivouacked in the sand-dunes, and the following day reached some wells known as Huwail, close to which was a single date-mat shelter occupied by an aged Badoo of the Bani Yas tribe, who made his livelihood by helping to water camels. He possessed a fine she-camel and foal, and on our arrival apparently announced his intention of killing the foal, as he said, 'in honour of Shaikh Zaid's friend'. I was having my tub when I got the message, and shouted out immediately that I did not want the meat and that the foal was not to be killed, but it was too late. The old man had stabbed it in the chest and cut its throat before I could stop him, and the wretched mother moaned restlessly for her offspring for the rest of the day. My followers were no doubt inwardly delighted, as they got a sumptuous meal while I had to make generous amends to the old man.

On both journeys he passed through Buraimi oasis, where Al Ain is today located. He pin-pointed its locality, which until then had been uncertain. He went to Ibri, which he found much improved since it had been castigated by Samuel Miles, and Nizwa.

As early as the 1920s, geologists were beginning to seek oil in Oman and the Trucial States, and they covered much of the area traversed by Miles and also employed camels (see Eccles 1927).

# GERTRUDE BELL

**Figure 3.32** Gertrude Bell in c.1910.
*Source: Encyclopaedia Britannica.*

Gertrude Lowthian Bell (1868–1926) (**Figure 3.32**) was a traveller, campaigner against female suffrage, archaeologist, diplomat and spy. She was born in Country Durham in 1868, and had access to considerable private means, as her family were owners of steel, aluminium, chemical and coal enterprises. It was said that she was born with an iron spoon in her mouth. She had a remarkable career at Lady Margaret Hall, Oxford, where, in spite of her prowess in a whole range of sports, she obtained a first class degree in modern history, the first woman to win that particular distinction at Oxford. She was slender, had piercing green eyes, a long pointed nose, a ramrod posture, and a mass of auburn hair. She was a lifelong cigarette smoker.

At the dawn of the twentieth century she started a series of travels and archaeological researches in Arabia and the Middle East, including a crossing of the Syrian Desert, which led to the publication of her classic *The Desert and the Sown* (1907). This book grew great acclaim, detailing as it did the journey of a lone Englishwoman, who was an accomplished scholar, linguist and writer. In it (p.1) she expressed the feelings that created her enthusiasm for desert travel:

> To those bred under an elaborate social order few such moments of exhilaration can come as that which stands at the threshold of wild travel. The gates of the enclosed garden are thrown open, the chain at the entrance of the sanctuary is lowered, with a wary glance to right and left you step forth, and behold! the immeasurable world. The world of adventure and of enterprise, dark with hurrying storms, glittering in raw sunlight, and unanswered question and an unanswerable doubt hidden in the fold of every hill. Into it you must go alone, separated from the troops of friends that walk the rose alleys, stripped of the purple and fine linen that impede the fighting arm, roofless, defenceless, without possessions.

In 1909 she surveyed the Roman and Byzantine fortresses on the banks of the Euphrates in Mesopotamia. From Aleppo, she travelled via the Hittite capital at Carchemish to Mesopotamia and Baghdad, before going upstream to Mosul and then crossing to Konia in Turkey. This was her most important journey of exploration, described in *Amurath to Amurath* (1911). During 1911 she made a further expedition to Mesopotamia.

In 1913–14 she travelled by camel to the walled city of Ha'il, an oasis in the heart of central Arabia and a stopping point on the pilgrimage from Mesopotamia to Mecca. She used scientific instruments throughout her journey, made observations of latitude, temperatures and pressures, and took excellent photographs with her expensive cameras, including panoramas, of many archaeological sites, some of which have now been eroded or pilfered. Indeed to her, scientific endeavour was more important than mere adventure.

Gertrude was never able to write a full account of her journey to Ha'il, and the task was left to D.G. Hogarth to put together a narrative of the journey after her death from diaries, letters and recollections of conversations (Hogarth 1927). Many warned her against the journey, not least because of grave political uncertainties involving the Turks and rivalries between Ibn Saud and Ibn Rashid. Among those who worried was her paramour, Charles 'Dick' Doughty-Wylie. Although some claim that he was braver on the battlefield than in bed, he and Gertrude were very close. He was awarded the VC but was killed in 1915. Gertrude was devastated.

In spite of the warnings, she set off from Damascus in December 1913. From the start, the journey was marred by bitter winds, torrental winds, a tribal attack and a contretemps with Turkish troops. It took her three months to reach Ha'il via the Nejd. Her caravan consisted of twenty camels, three camel drivers and three other escorts. She was more or less held prisoner in Ha'il, and decided not to proceed further to meet Ibn Saud. Instead, she resolved to travel to Baghdad via Karbala. The whole journey to and from Damascus took four and a half months. Gertrude regarded her journey as a failure. As Douglas Carruthers wrote (1958, 53), 'She never met Ibn Rashid, far less Ibn Saud, and only touched the fringe of Najd proper. In consequence, she was bitterly disappointed and did not take the failure kindly. She wrote not one word, when she might have written an epic, even if it was of a failure.'

Bell's mode of travel in the desert was relatively lavish, as described by Howell (2006, 131):

> She packed couture evening dresses, lawn blouses and linen riding skirts, cotton shirts and fur coats, sweaters and scarves, canvas and leather

> boots. Beneath layers of lacy petticoats she hid guns, cameras and film, and wrapped up many pairs of binoculars and pistols as gifts for the more important sheikhs. She carried hats, veils, parasols, lavender soap, Egyptian cigarettes in a silver case, insect powder, maps, books, a Wedgwood dinner service, silver candlesticks and hairbrushes, crystal glasses, linen and blankets, folding tables and a comfortable chair – as well as her travelling canvas bed and bath.

Howell (2006, 117) has summarised the magnitude of Bell's 1911 and 1913 expeditions:

> Taken together, her journeys encompassed most of Syria, Turkey and Mesopotamia . . . She covered more than ten thousand miles on the map, but she went over hills and mountains, searched for fords, took detours to ancient sites and to make contact with sheiks. Over six hundred days she must have journeyed at least 20,000 miles in the saddle.

She was awarded the RGS Founder's Medal in 1918, an honour only twice previously conferred on a woman.

In 1915, after the outbreak of the First World War and at the instigation of D.G. Hogarth, she joined the military intelligence department in Cairo, the aim of which was to hasten the departure of the Turks from Arabia and to develop links with the Arabs that would strengthen British influence once the war was over. Among those with whom she worked was T.E. Lawrence. In 1916 she went to Basra, working on the staff of Sir Percy Cox. She became a high-ranking colonial administrator, and was a crucial person in the establishment of the modern state of Iraq (and of its National Museum in Baghdad) and in the assumption of the Iraqi throne by King Faisal.

In 1926, unlucky in love (and not for the first time), she died in the Iraqi capital, suffering from depression and having taken an overdose of sleeping tablets. Some ascribed her death to natural causes, and her obituary in *The Times* (13 July, 1926) suggested that it was due to overwork and 'an unflinching sense of duty which, except for two very short holidays at home, kept her for nearly ten years almost continuously at Baghdad, even during the torrid heat of the Mesopotamian summers, when the thermometer stands for weeks together about and above 120 deg. in the shade'. She had lived life to the full in a man's world. Her friend David Hogarth penned a fulsome obituary of her in the *Geographical Journal* (1926, 363), remarking:

> No other woman of recent time has combined her qualities – her taste for arduous and dangerous adventure with her scientific interest and knowledge, her competence in archaeology and art, her distinguished literary gift, her sympathy for all sorts and conditions of men, her political insight and appreciation of human values, her masculine vigour, hard common sense and practical efficiency – all tempered by feminine charm and a most romantic spirit.

Her archives are largely collected in the University of Newcastle.

## 'JACK' PHILBY

Harry St John Bridger Philby (1885–1960) (**Figure 3.33**), generally known as Jack, was born in Ceylon, the son of a coffee grower. He was one of three great British explorers of the Great Southern Desert of Arabia, which is generally known as the Rub Al' Khali or Empty Quarter. The other two were Bertram Thomas and Wilfred Thesiger. Educated at Westminster School and Cambridge University, he excelled at languages and was accepted into the Indian Civil Service. At his first marriage in 1910 his best man was a distant cousin and young soldier, Lieutenant Bernard Montgomery.

Various people had already travelled to the Nejd by camel, including Douglas Carruthers (1910), but this was something that Philby was also determined to do. In 1917 he contrived to go as Head of Mission to Ibn Sa'ud, ruler of the Nejd in central Arabia, with the hope of seeking his support in the war against the Turks. He travelled with a small group by camel from the Arabian Gulf to Riyadh and then across to Jedda on the Red Sea (*Heart of Arabia*, 1922). On this coast-to-coast trip he developed an undying admiration for Ibn Sa'ud. He also met the Hashemite ruler of the Hejaz, Sharif Husain, whom he thought less suitable as a British ally. In 1917–18, he explored the southern province of the Nejd. For this and earlier journeys he was awarded the RGS Founder's Medal in 1920. He reported the journey in the *Geographical Journal* for 1920, and besides giving descriptions of the tribes, the settlements, irrigation systems and a wadi flood, he used it as an occasion to cast doubt on the veracity of Palgrave's accounts.

In 1922, Philby joined up with Major A.L. Holt, who had been investigating the possibilities of railway construction in Arabia, on an expedition across from Amman to Jauf and on to Karbala on the Mesopotamian border. What was significant about this expedition was that for part of the journey it used a small fleet of Ford cars of the type that Ball and Williams had utilised in Egypt in the

**Figure 3.33** Philby in Arab attire. *Source: Philby 1922, volume 1, 38.*

First World War, though Philby maintained a personal preference for camels (Philby 1923). This did not stop him, however, from using cars in the 1930s to explore Arabia and to drive to and from England.

Never an easy man, being obstinate and combative and seeing everything in terms of black and white, Philby left the public service in 1925 and founded a trading company in Jeddah, where he became a close friend and adviser of Ibn Sa'ud. In 1930 he converted to Islam and was known as Sheikh Abdullah by the Saudis. He made the hajj, or pilgrimage, to Mecca in 1931, riding with the King and in the company of 10,000 camels (Philby 1946).

In the 1930s Philby travelled in the borderlands of the Yemen but is now known most of all for his attempt to be the first man to cross the Rub' al Khali. In the event he was beaten by Bertram Thomas, who, without seeking the permission that delayed Philby, crossed from south to north in 1931. However, it can be argued that Philby studied it more thoroughly. Notwithstanding his profound disappointment that 'The virgin charm of the Empty Quarter had been unveiled by another', Philby set out in January 1932 with food for three months of meandering travel and with 32 camels to carry his men and supplies. The camels, with one exception, were all female, and were various strains of the famous 'Umaniya breed. He travelled from north to south over a distance of 2700 km. His journey, upon which he came across the meteorite impact craters of Wabar, was described in *The Empty Quarter*, which was published in 1933.

At Salwa creek the camels proved to be perverse (p.48):

> At times the water at this ford is deep enough to necessitate a wide circuit to the southward but we found it barely knee-deep. Yet the camels had to be coaxed, forced or led into, the water, and such is the perversity of their race that no sooner had they reached mid-stream than some made vigorous efforts to bathe while others, having refused to drink fresh water at Ba'aij stretched forth their long necks to take a gulp of the salt water as they passed. It is not good for them and the Arabs showed no sympathy for such depravity.

In some areas the camels had problems with gravel surfaces (p.145):

> The sharp stones ... tore the tender pads of the sand-bred animals to ribbons and they marched daubing the desert with red spots of blood. The Arabs' treatment for such troubles consists of simple botching. A strip or round of leather is neatly cut out to the required size ... and soaked with butter which is also liberally smeared over the pad to be doctored. Strong men then hold down the struggling beast on its side while the cobbler ... threads a stout bodkin with a thin leather thong and draws it through patch and sole, repeating the process until the former becomes more or less part of the latter. Rough ends are then cut or smoothed off, and camel rises, apparently none the worse for the operation, to face more bravely a new world of gravel or grit.

Thirst was a more general issue, and in some cases the camels were force fed with water – a process called 'snuffing' (p.267):

> This is an economical method of refreshing camels by administering a kettleful of water through the nostrils to cool the head and brain. The silly beasts struggle violently under such treatment, which is presumably painful enough at the time, but there would seem to be no doubt about its efficacy . . .

Another problem was that camels could be flatulent. To cure this (p.276):

> . . . The Arabs resort to a curious device. They sew up the anal orifice or plaster it over with a dough of camel dung, and then tie down the tail, fastening it to the saddle with a cord passed under its belly and between the legs, so as to prevent the emission of wind.

In his book Philby spent some time (p.326 et seq.) in comparing the speed and endurance of his expedition in comparison with that undertaken by Gerhard Rohlfs in the Libyan Desert some 60 or so years earlier. Wilfred Thesiger, in his obituary of Philby in the *Geographical Journal* for 1960, argued that Philby's journey was a greater one than that of Bertram Thomas, and that it was 'The greatest in the story of Arabian exploration'. Thesiger also spoke appreciatively of Philby's important collections of birds, mammals, plants and insects, and his scholarly work on ruins and inscriptions. He also found indications that the desert had once been better watered, causing rivers and human populations to expand, possibly in Neolithic times. He experienced the phenomenon of singing sands in the Rub' al Khali dunes, and described it as being like 'a siren or perhaps an aeroplane engine – quite a musical, pleasing, rhythmic sound of astonishing depth'. He also discovered fulgurites – sticks of sand that had been fused by lightning strikes.

During the Second World War Philby was a pacifist and appeaser, attacked British policy (for which he was briefly imprisoned) and spent most of the remainder of his life in Saudi Arabia (where he purchased a second wife) and Lebanon, where he died in 1960. At the time of his death he was staying with one of his sons, Kim, who gained notoriety as a Soviet spy during the Cold War. Overall, Philby may have been an unpleasant individual and J.B. Kelly (1977) listed a series of his failings that included being a prig at school, a muddled communist, an admirer of Hitler, a philanderer, a bigamist, somebody who was negligent, capricious and moody towards his family, and a person whose life was dominated by two obsessions – the making of money and the winning of fame.

Figure 3.34 Bertram Thomas astride a camel. *Source: Thomas 1931, frontispiece.*

## BERTRAM THOMAS

Bertram Thomas (1892–1950) (**Figure 3.34**), born in Bristol, was educated at Cambridge University. From Cambridge he went to Flanders, and was invalided back home in 1915 after seeing action at both Mons and Ypres. He came to know the Arab world through his army service from 1916 onwards in the Somerset Light Infantry. He worked as an administrator in Trans-Jordan and Oman, where he was financial adviser to the Sultan and given the title of *wazir*. He was the first European to cross the Rub' al Khali, a journey he described in *Arabia Felix* (1932) and for which he was awarded the RGS Founder's Medal. The book was a bestseller in its day and far superior to Philby's rather turgid account of his own crossing of the Rub' al Khali.

Thomas outlined his main motive for this pioneering journey, which started from Muscat on 4 October, 1930 (p. xxiv):

> The remote recesses of the earth, Arctic and Antarctic, the sources of the Amazon, and the vast inner spaces of Asia and Africa, have one by one yielded their secrets to man's curiosity, until by a strange chance the Rub' al Khali remained almost the last considerable *terra incognita* . . .

Without official sanction, Thomas set out from Muscat by sea and went to Salalah, a settlement on the southern coast of Oman. With a caravan of 30 men and 40 camels, he suffered the usual deprivations of desert travel and was forever wary of tribal raids (**Figure 3.35**). He crossed the great dunes of the Uruq-Adh-Dhahiya (p. 170):

> Very impressive is a great dune region at first sight – a vast ocean of billowing sands, here tilted into sudden frowning heights, and there falling to gentle valleys merciful for camels, though without a scrap of verdure in view. Dunes of all sizes, unsymmetrical in relation to one another, but with the exquisite roundness of a girl's breast, rise tier upon tier like a mighty mountain system.

The dunes were hard going, but they were also beautiful (p.174):

> Our toil had its compensations. There were moments when we came suddenly upon a picture of sublime grandeur, an immense and noble plastic architecture, an exquisite purity of colour, old rose-red, under the cloudless sky and brilliant light.

Thomas had much to say about his camels. For example, he discussed the camels of the Qara Mountains (p.79):

> The camel is bred in the wooded valleys below on a fairly extensive scale, but by limited groups, for its rewards to man are less and its care makes greater demands; even the sexual act necessitates man's assistance. Milch camels normally never know a saddle, and miserable mounts they make, as I know by costly experience. Their breeding is only for the production of he-camels, the stronger sex, to serve the transport needs of the province, and carry frankincense from the mountains to the coast, and to return with sardines to the mountains for cattle fodder.

**Figure 3.35** Bertram Thomas (centre) with some of his men. Thomas was tall – 188 cm.
*Source: Thomas 1932.*

He noted how his men looked after the chiropody of their camels (p.115):

> 'Come and see ahjat an naga,' shouted Shaikh Salih, who sat on the ground with a large sail-maker's needle in his hand, bending over a camel that was held down on her side by a few Arabs in the customary way; one held the head turned back along the body; another had a twitch on her upper lip, a third held one leg stretched out behind. The operation was to sew a small leather patch on the side of the foot over an abrasion which had been caused by the unaccustomed stony tracks of the mountains; it would be a protection on the morrow's journey.

Bertram discussed the sex act of camels and, as Buffon had done c.150 years previously, mentioned the way in which humans were involved (p.214):

> Camels are thus, like the llama and the lion, rare in the animal kingdom in the performance of the act in a sitting position. The Badu master is necessary to the operation, scooping the sands round the cow's legs for her comfort, inserting the penis – the formation of which is in reverse axis to nearly all the rest of mammal creation, and interfering after a few minutes to drive the bull off. After ten days if no result is apparent, the cow's master will find another bull to serve her. The sign of pregnancy is the flag-wagging of her ridiculous tail when approached by a rider to mount.

He discussed the relative value of male and female camels (p.267):

> The prosperity of the tribe is measured by the number and condition of its camels. The sources of wealth are good pastures and the manly prowess of its members who will aggressively acquire fresh camels at their enemies' expense. Camels therefore fall into two classes, herds of milch camels (the assets and reserve) and the less numerous riding camels (working capital). The first may be worth one hundred dollars each, the second from two hundred to four hundred for an exceptional animal. The milch camel never knows a saddle, and is raised solely for breeding and to produce milk and wool. The female is therefore the valued sex, and the cow calf is always reared, whereas the bull calf is a luxury not worth his keep. In consequence he seldom survives the first year of life, and not infrequently is slaughtered for food on the day he is born without his mother seeing him. Normally two or three bulls will

> serve a herd of fifty cows; they also carry tents and the heavier burdens when on the move.

Finally, he described the ways in which camels are controlled and how they are ridden (p. 286):

> Mounted, she is always given her head, the halter being frequently unused, only a tapping on her neck with the cane, and a few guttural noises being required to teach her to obey her rider's will. Her own disposition tempts her to almost every tuft of food in the way; even she will crunch some fragment of white desiccated bone against her ridiculous toothless upper jaw unless her master urges her past in discouragement, as he normally will. And so the Badu sits jogging along hour after hour with an occasional change of seat. In an Omani saddle he will not ride astride for long, but tuck his feet up under his haunches in a sitting-kneeling position, or ride side-saddle with his legs dangling limply.
>
> The double-poled Murra saddle admits of less variety, but is more comfortable to Europeans.

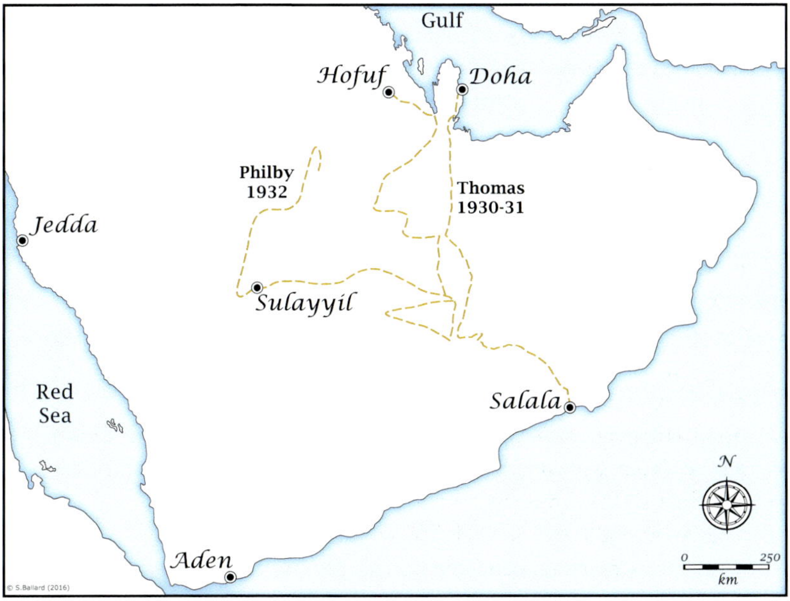


**Figure 3.36** Thomas's and Philby's routes across the Empty Quarter. *Source: Goudie 2016.*

Thomas reached the northern edge of the desert near Doha on 28 January, 1931 (**Figure 3.36**). A few days later he saw the waters of the Persian Gulf. He had stolen the prize of making the first crossing of the Rub' al Khali. Jack Philby had to take second place. His expedition was notable for the excellent notes he took, and for the specimens he collected. Subsequently, Thomas worked in Lebanon as first director of the Middle East Centre for Arab Studies, and then as Shell's head of operations in the Gulf. He died in 1950, alcoholic and overweight (according to William Atkins, 2018, 20) at the age of 57. Sir Percy Sykes (1933, 306), himself a great explorer in Persia, in summing up Bertram's achievements, picks out his physical fitness and hardiness and concluded that 'No living explorer has prepared himself more thoroughly for a great task and none has achieved greater success'.

Another person who passed through Arabia at this time was W.B. Seabrook (1884–1945). An American, he was a strange individual. He had deep interests in voodoo and cannibalism. He visited Arabia in the 1924 and penned a book, *Adventures in Arabia* (1928). He was scornful of the baggage camel (p.55):

> He is ugly as sin, and he does not belie his appearance. His black heart is filled with a vicious and melancholy hatred. He frequently stinks, or rather his breath does, like ten thousand constipated devils. And one of his favorite tricks is to vomit his vile greenish cud explosively in the face of the man who may be seeking to do him a kindness. He will bite you, not in anger as high-tempered horses sometimes do, but in casual, cold contumacy. If you are lying asleep, he will walk out of his way to step on your face . . . He is lazy and treacherous. Nobody loves him – not even Allah, the Merciful, the Compassionate.

On the other hand he had much more time for another breed of camel (p.56):

> But there is another breed of camel in the Arabian desert, different from these sons of Shaitan, as a pure-bred greyhound is different from a hyena – the white hejin, or racing camels, the pride and glory of their owners.

These he did find 'superbly beautiful animals, pure white, long and clean of limb, with graceful necks, and small, well-formed heads'.

## WILFRED THESIGER

Although he was something of an anachronism, many observers have regarded Wilfred Thesiger (1910–2003) as the greatest desert explorer of the twentieth century. Whether this is true is debatable, but he was certainly a great traveller, writer, circumciser and photographer. He was born in Addis Ababa. Thesiger surmised that the 'perverse necessity' that drove him later in life to the deserts of the East may have lain in his childhood in Abyssinia, and distant memories of such things as the smell of dust and acacias under a hot sun, and of camel herds at water holes. After attending prep school in Sussex, he was sent to Eton and then to Oxford to read history and to box. Soon after going down from Oxford and only 23 years old, he traversed Danakil in 1933–1934. His aim was to explore the Awash River and the Aussa sultanate with its Danakil nomads, the Dankali, who were suspected of a tendency to kill men and carry off their testicles as trophies (Thesiger 1935). Such practices, however, probably held little horror for Thesiger, who had survived fagging and flogging at school. He started out with 18 load-carrying camels and with 40 Abyssinians and Somalis, all armed with rifles.

He described the Danakil people he encountered (p.2–3):

> They have a ready smile, and though many of them file their teeth their expression is attractive. The boys have their hair shaven off or cut very short, but the men allow it to grow into a fuzzy mop. They anoint it constantly with ghee, and periodically wave it into close ringlets. The women, and occasionally the men, raise scar-patterns upon their bodies, and both bear the tribal mark upon their faces. The men wear short and usually extremely dirty loin cloths, and they keep themselves carefully covered except when working in the wells. The women wear long skirts dyed brown in colour with the bark of the mimosa. Their breasts are bare, but if married they wear a piece of black cloth, known as *shash*, round their heads. They are fond of beads, but do not weave them into their hair as do the Arussi women. The men invariably wear strapped across the front of the stomach a formidable curved knife known as a *jile*, with a blade 16 inches long and sharp on both sides.

He later expanded this discussion of the Danakil people (1960, 10):

> Among them a man's standing depended to a very large extent on his reputation as a warrior, which was judged by the number of men he had

> killed and mutilated. There was no need to kill another man in fair fight; all that was required to establish a reputation was to collect the necessary number of severed genitals. Each kill entitled the warrior to wear some distinctive ornament, and ostrich feather or comb in his hair, an ear-ring, bracelet, or coloured loin cloth.

It was on this six month long journey from Awash Station to Tajura that he seems to have developed his interest in circumcision, something at which he became a master in later life when he travelled in Iraq.

After Danakil in 1935, Thesiger joined the Sudan Political Service and travelled to Tibesti in the heart of the Sahara. On that journey he discovered the motivation that drove him (Thesiger 1935, 22):

> In the desert I found a freedom unattainable in civilization; a life unhampered by possessions, since everything that was not a necessity was an encumbrance. I had found, too, a comradeship inherent in the circumstances, and the belief that tranquillity was to be found there. I had learnt the satisfaction which comes from hardship and the pleasure which springs from abstinence.

This was a journey carried out by camel (Thesiger, 1939) and during this journey of three months duration he covered some 3200 km in all, and climbed to the highest summit at 3391 m above sea-level. He explained his motivation (p.434)

> The wish to visit Tibesti had grown strongly on me during the three years that I had served in northern Darfur, and I decided to spend my 1938 leave in travelling among these mountains. I left Tini on the Darfur-Ouadai frontier on 3 August 1938. I had hired camels to take me as far as Borkou, after which it would be necessary to use Tibbu camels accustomed to travelling among the mountains. Besides the Zaghawa who accompanied these camels, I had with me Idris Daud, a Zaghawi lad of the Kobe tribe who had been with me for the past three years, and an elderly Bideyi named Khatir, who knew the Tibbu tongue, having lived among these tribes as a boy. We travelled light, the distances before us being great and the time at our disposal short. We carried flour, sufficient to last us for three months, from which to make the assida or flour paste on which we lived, dried meat, onions, and ladies fingers . . . for the mulah or seasoning, and a sufficiency of sugar

> and tea. We had no kit other than our rifles, sheepskins, blankets, and a small tent.

Thesiger encountered French troops (p.436):

> Two sections of the Groupe Nomade or camel corps passed through Fada the day that I arrived, and I was very pleased to have the opportunity of seeing this splendid body of men. These Groupes Nomades are composed of Tirailleurs Senegalais commanded by French officers. They live continuously under active service conditions and are in consequence extraordinarily fit and hard. They remain far out in the desert, moving from place to place continuously. Few places are unvisited by them, though their movements naturally depend to a certain extent on the grazing. They delight especially in choosing inaccessible and waterless spots for liaisons with neighbouring Groupes Nomades. The life of these officers is a hard one. It takes three months to arrive from France. They then serve for two years, with the choice of extending the period for a third year, before a visit to France and transfer elsewhere. The Groupes Nomades trek in a country harsh and bare in the extreme, and though they carry tents they have no camp kit other than a folding office table and chair.

However, during his service in Sudan he spent most of his time in the dry bush of Darfur, northwestern Sudan. It was in Darfur that he first learned to travel by fast-riding camel with local companions, dressing as they did, eating local food out of a shared bowl, and asking nothing of technology but a torch, a compass, and above all, a decent rifle. Employing the finest riding camels, he once rode 184 km in 23 hours, and rode from Jabal Maidob to Omdurman, a distance of 720 km, in 9 days.

He had further desert experience in the Second World War, when he fought with the Special Air Service in the Western Desert of Egypt. But here he felt that he was insulated from the desert by travelling in a Jeep, something he regarded as an abomination. However, he considered all this desert experience as a mere prelude to the five years of desert travel for which he is most famed – in the Empty Quarter of Arabia. This was made possible when in Addis Ababa, towards the end of the war, he met O.B. Lean, the Desert Locust Specialist of FAO, who was looking for someone to travel in southern Arabia to collect information on locust movements. It provided Thesiger with the political permissions and the opportunity to enter an otherwise inaccessible land. As he wrote (1987, 397):

**Figure 3.37** The surface of the great inland Sebkha of Oman – the Umm al Samim. *Source: Author.*

> Only in Arabia did an enormous tract of desert, which even Arabs call Rub' al Khali, the Empty Quarter, remain largely inviolate, offering the final challenge of desert exploration.
>
> Thomas and Philby had crossed the Empty Quarter in the early 1930s but no other European had been there since, no aeroplane had even flown over it, and no car had been nearer than the RAF camp at Salala, on the shore of the Indian Ocean, or the townships on the Trucial Coast. In contrast with the Sahara, vast areas of the Empty Quarter were still unexplored and it was surrounded by a no-man's-land of warring tribes.

Thesiger left Addis in March 1945 and stayed in Arabia, apart from brief periods of leave, until 1950. He made a series of journeys by camel to Dhaufar and the Hadramaut, to the southern Hejaz and to the Trucial Coast and Oman, where he described the quicksands of the Umm al Samim (**Figure 3.37**) and the Wahiba Sands (**Figure 3.38**). But, most importantly of all, he made two crossings of the Empty Quarter: one in 1946–7, and the other in

1948. The first of these journeys took him from Salala on the Arabian Sea across the Rub' al Khali to Liwa, then back to Salala by the steppes of western Oman and the Jaddat al Harasis, and from there through the northern Mahra country to Mukala. He watched with care how the Bedu mounted their camels (1960, 37):

> A Bedu who is going to mount a couched camel stands behind her tail. He then leans forward and catches the wooden tree with his left hand as he places his left knee in the saddle. Immediately the camel feels his weight she starts to rise, lifting her hindquarters off the ground, and he swings his right leg over the saddle. The camel then rises to her knees and with another jerk is on her feet. The Bedu either sit with a leg on either side of the hump, or kneel in the saddle, sitting on the upturned soles of their feet, in which case they are riding entirely by balance. They prefer to ride kneeling, especially if they mean to gallop.

Thesiger did not emulate them, finding that kneeling was too uncomfortable and too precarious. He remarked that he had never seen a Bedu strike or

**Figure 3.38** Camels in the great linear dunes of Wahiba, Oman. *Source: Author.*

ill-treat a camel (p.58), had seen them 'fondling and kissing them whilst they murmured endearments'. He recounted how one camel was so attached to her owner that at intervals through the night she came over to him as he slept, moaning softly at him where he lay, before going back to graze. He also reported, as had others, that the Bedu had to assist camels to mate as otherwise they would 'never get it in the right place' (p.101).

The second journey, from January to March, 1948, took him to more westerly parts of the sand sea. On these journeys with his Bedu companions, he enjoyed various privations (Thesiger 1987, 398):

> ...We endured almost incessant hunger and, worse still, thirst, sometimes for days on end rationing ourselves to a pint a day; there was the heat of a blazing sun in a shadeless land; the bitter cold of winter nights; incessant watchfulness for raiders, our rifles always at hand; anxiety that our camels, on which our lives depended, would collapse.

The compensations were the clean beauty of 'the sculptured shape of dunes that rose to seven hundred feet and extended for a hundred miles', and the generosity, honesty, pride, beauty, and loyalty of his young companions, notably the teenage Salim bin Kabina and Salim bin Ghabaisha (**Figure 3.39**). They both played a pivotal role in Thesiger's exploits from 1945 onwards and in *Arabian Sands* he provides a list of the journeys with which they were involved. The book is also dedicated to them. Thesiger's black and white photographs of both the landscapes and these people are breathtaking.

Thesiger regretted the modern world. As he lamented in the Introduction to *Arabian Sands* (1959, p. xiii):

**Figure 3.39** Salim bin Ghabaisha (left) and Salim bin Kabina (right), standing on a rock outcrop in Oman in c.1950. Each man carries a rifle, and wears a *khanjar* (dagger) on a cartridge belt over his *thobe* (long shirt). *Source: Thesiger 1959, figure 68.*

> I went to Southern Arabia only just in time. Others will go there to study geology and archaeology, the birds and plants and animals, even to study the Arabs themselves, but they will move about in cars and will keep in touch with the outside world by wireless. They will bring back results far more interesting than mine, but they will never now the spirit of the land nor the greatness of the Arabs . . . Today the desert where I travelled is scarred with the tracks of lorries and littered with discarded junk imported from Europe and America.

After Arabia, Thesiger continued his nomadic life, travelling *inter alia* in Iraq, Pakistan, Afghanistan, Morocco, Ladakh, Ethiopia and northern Kenya. Knighted in 1995, and recipient of the RGS Founder's Medal in 1948, he died in Surrey aged 93 in 2003.

# CHAPTER FOUR

# EPILOGUE

## VIEWS ON CAMELS

One matter that arises from reading the writings of many of the authors cited in this book is that over the centuries humans have offered diverse and divergent views on these extraordinary animals. At one extreme is the view that camels are stupid and malicious brutes, while at the other, they are seen to be intelligent, patient, docile, indispensable, and even loving friends of humans. The following quotations from the text bring this out (**Table 4.1**):

Table 4.1. Views on camels

**Sir Samuel Baker**
Far from being the docile and patient animal generally described, it is quite the reverse, and the males are frequently dangerous. They are exceedingly perverse; and are, as before described, excessively stupid.

**Charles Belgrave**
They are not affectionate animals and they never seem to know their own masters; there was only one among mine that had any 'parlour tricks', and he used to inhale tobacco smoke through his nose, apparently with the greatest appreciation.

**Anne Blunt**
The camel, whatever his faults, is certainly not ill-tempered, and his roaring is as little terrible to any but cockney ears as the lowing of a cow. Roaring is his manner of speech and need frighten no-one.

**Angus Buchanan**
The awkward and somewhat wooden appearance of camels does not lead one to associate much intelligence with them, but to think so is a mistake, and if one desires to have a really good hunting camel, I know of no better method to secure it than to select a good-natured beast from the rank and file, and hand feed it with tit-bits of vegetation, and pet it when mounting and dismounting, and let no one else saddle it or ride it, and before long you will be astonished to find that you have won a queer pet and a useful and obedient comrade.

**Charles Furlong**
. . . on the whole, this ancient burden-bearer of man is a dignified, long-suffering, lugubrious anomaly. He is not aggressive; his indifference toward man seems almost contemptuous. He is imperturbable and patient beyond precedent, and on the march will continue to stagger on until his last ounce of strength has been exhausted.

**Ahmed Hassanein**
The qualities of a camel are seldom, if ever, appreciated on a slight acquaintance. The camel is as clever as a horse, if not more clever, and in some ways is more human.

**Jack Keane**
This huge ungainly brute is endowed with a stupidity and helplessness perfectly pitiable, and its temper is diabolical. While being laden it gives vent to spuming, spluttering, bellows and whines not unlike what the roars of an enraged croupous lion might be supposed to be. I never recall to my mind a camel, but I picture it biting, or trying to bite something, an operation for which its peculiar dental arrangement well adapts it.

**Ludovico Nesbitt**
The longer you associate with your animals, the more human they seem to become, especially when the association is as close as ours was on this expedition.

**Gifford Palgrave**
I have, while in England, heard and read more than once of the 'docile camel'. If 'docile' means stupid, well and good; in such a case the camel is the very model of docility. But if the epithet is intended to designate an animal that takes an interest in its rider so far as a beast can, that in some way understands his intentions or shares them in a subordinate fashion, that obeys from a sort of submissive or half fellow-feeling with its master, like the horse and elephant, then I say that the camel is by no means docile, very much the contrary . . .

**W.B. Seabrook**
He is ugly as sin, and he does not belie his appearance. His black heart is filled with a vicious and melancholy hatred. He frequently stinks, or rather his breath does, like ten thousand constipated devils. And one of his favorite tricks is to vomit his vile greenish cud explosively in the face of the man who may be seeking to do him a kindness.
He will bite you, not in anger as high-tempered horses sometimes do, but in casual, cold contumacy.

**Hanns Vischer**
The camel I rode from Bilma to Bornu soon became attached to me, like a dog, and would follow me about everywhere . . . Many a night, to guard against the piercing wind, I slept alongside my good warm animal, which took me to his bosom like a mother.

**James Wellsted**
Notwithstanding its patience and other admirable qualities, the camel is gifted with but little sagacity, nor does it appear to be capable of forming any strong attachment to its master, although they frequently do so to one of their own kind with which they have long been accustomed to travel . . .

## CAMEL CATASTROPHES

Another matter arising from the experiences related in this book, is the propensity of some camels to expire rather unexpectedly (**Figure 4.1**). This is shown in **Table 4.2**.

**Figure 4.1** *Shipwreck in the Desert* by Carl Haag (1886). *Source: Author.*

Table 4.2. Camel catastrophes

| |
|---|
| **Angus Buchanan**<br>We find that he has collapsed wearily in the sand and does not want to move. He is far through, but we cannot camp and wait beside him. So in a little time he is persuaded to rise to his feet again; and the caravan moves slowly on. But it is not very long before the old fellow gives up again, for his is a losing fight in the full heat of the midday sun. We try for a little to encourage him to get up, but to no avail. He is past further struggle. I order the caravan to move onward, and remain behind with Elatu, beside the prostrate animal, for I cannot leave the poor brute to die a slow, lingering death, with the agony of pitiless surroundings holding finality immediately before his eyes. When the caravan is distant there is a single revolver shot, and we have one less in our band. |
| **Ahmed Hassanein**<br>Camels, the best of them, and for no apparent reason, give in, as it happened when I left Kufra and one of my best camels died on the second night; while, on the other hand, the weakest camel of the caravan, which left Kufra tottering under its load, went through the whole trip, about 950 miles, and arrived tottering at El Fasher. 'God will protect it', said its Beduin owner when rebuked for bringing such a sorry animals, and in truth God did protect it. |
| **Ludovico Nesbitt**<br>. . . one of our camels cast himself down, and had to be abandoned. . . .It was one of the last we had purchased atbthe outset, at Awash Station. It fell into one the subsided domes, and could not rise again. After giving the faithful beast a farewell caress, I took a rifle from Wold Jesus and shot it. |
| **Joseph Pitts**<br>If it happen that the camel dies by the way, the carrier is to supply us with another; and therefore, those carriers who come from Egypt to Mecca with the Caravan, bring with them several spare camels; for there is hardly a night passeth but many die upon the road, for if a camel should chance to fall, it is seldom known that it is able to rise again; and if it should, they despair of its being capable of performing the journey, or ever being useful more. It is a common thing, therefore, when a camel once falls, to take off its burden and put it on another, and then kill it; which the poorer sort of the company eat. |

In addition, as shown by the appalling carnage during the march of the Army of the Indus, and by the experience of some French military forays in the Sahara, badly treated or inexperiencedly handled camels can and did die in their thousands.

## THE SPEED, RANGE AND LOAD-CARRYING CAPACITY OF CAMELS

One of the remarkable features of the camel that becomes apparent from all the descriptions of the great explorers is just how capable camels could be of moving fast, travelling huge distances, and carrying large loads.

T.E. Lawrence wrote a short but informative paper on the speed at which camels can travel (1919, 71):

> For speed, the best performance I know was the 39 hours' ride of Sherif Barakat ibn Smeiyah, from Medina to Mecca, by the Rabegh road a few years ago. It was a race, and camels were changed at Rabegh, 154 miles from Medina. The total distance works out at about 280 miles, and it was covered practically without a stop, except for a few minutes at Rabegh. The average speed was thus over seven miles per hour. A race of this sort is a test of the man's endurance, rather than that of the camel. Another equally fine ride was that of Aissa, a Harb tribesman, who came from Zilfi in Qasim to Yenbo in three days, and returned to Zilfi in four more, making a total of seven days for the round trip of 900 miles, an average of 130 miles a day. Aissa used four camels.
>
> Rides on single camels are more interesting as records. One of the Atram family of Fitenna Abu Tayi Howeitat, on a home-bred pedigree camel, rode between sunset and sunset from Nebk abu Gasr to Bair and Jefer, a distance of 143 miles. He rested in Jefer one day and returned on the third day to Nebk on the same camel. I owned this camel some years later, and found by experience that it would keep up a comfortable and steady trot of seven and a half miles per hour for hour after hour without special urging: but I never had need to do a trotting journey of greater length than from Rum to Akaba (39 miles) on it. It did this in a little under five hours, carrying a good deal of kit besides myself. Mesnid, a Sherari, took a message from Jefer to Akaba for me. He left Jefer at noon, and returned with the reply two days later at noon, doing 220 miles, and his errand also, in the 48 hours. He rode a Sherari four-year-old.
>
> The fast time for a camel-postman from Medina to Mecca is three days. This is an average of about 95 miles a day.
>
> With one servant I rode from Azrak through Jefer, Shedia, and Rum to Akaba (290 miles) in three days and a half. This is an average of about 84 miles a day. We rode Beni Sakhr camels.

> One of the Harith Sherifs of Modhig rode from Akaba to Mecca in nine days. The total distance is about 690 miles, which gives him an average of 77 miles a day. He rode a Sherari camel, and the trip is one of the finest I have heard of.
>
> A stripped camel, racing, will do something like twenty miles an hour for nearly two hours. For a short burst I have timed them trotting at 22 and cantering at 26 miles per hour.

Employing the finest riding camels, Wilfred Thesiger once rode 184 km in 23 hours in Darfur, and rode from Jabal Maidob to Omdurman, a distance of 720 km, in 9 days.

In general it seems that the best racing camels can manage speeds of up to 65 km per hour for short periods, and up to 40 km per hour for up to an hour. However, baggage camels do not attain such speeds and most caravans only tend to travel at between 3 and 5 km per hour.

As to range, this very much depends on availability of food, and most particularly, of water. If rains are good, so that moisture-rich feed is abundant, the range of camels can be impressive. George Murray (1926) mentioned that in the Eastern Desert of Egypt, following rain, camels lasted more than 140 days without being watered. He reckoned, however, that Ahmed Hassanein marched in ten days during May from Owenat to Erdi, a distance of c.450 kilometres, about the best that could be done without either water or grazing. In 1927, Newbold and Shaw did 440 km without watering from Tamar el Gusar to Selima, ten days' hard marching, having done another waterless stretch of c.380 km, nine days' march, a fortnight previously.

With regard to loads, Newbold and Shaw found that some of their camels carried 217 kg. for just under 1000 km. However, the average load was probably c.160 kg., but this lessened as water and grain were consumed. Wellsted gave a figure of 113–226 kg. Buffon suggested that some camels carried loads as great as 540 kg. This is similar to the load carried by a small pick-up truck, ten times that carried by the average donkey, and five times that carried by the average mule.

## PREMATURE DEATHS

Finally, it is worth recalling that some of the great explorers that have been discussed in this book died tragically young, especially in North Africa, in the nineteenth century:

William George Browne (1768–1813)
Friedrich Hornemann (1772–1801?)
Dixon Denham (1786–1828)
Hugh Clapperton (1788–1827)
Walter Oudney (1790–1824)
Jean-Raimond Pacho (1794–1829)
George Lyon (1795–1832)
René Caillié (1799–1838)
Richard Lander (1804–1834)
James Richardson (1806–1851)
Heinrich Barth (1821–1865)
Eduard Vogel (1829–1856)
Alexandrine Tinné (1835–1869)
Camille Douls (1864–1889)
Ludovico (Lewis) Mariano Nesbitt (1891–1935)
Ulrich Jasper Seetzen (1767–1811)
Johann Burckhardt (1784–1817)
James Wellsted (1805–1842)
Werner Munzinger (1832–1875)
David Carnegie (1871–1900)
Robert Burke (1821–1861)
William Wills (1834–1861)

To end on a more cheerful note, others, by having remarkable constitutions, by living at a time of progress in medicine, and by avoiding being murdered, were more fortunate, and lived to a great age, most notably Ardito Desio (1887–2001) who died aged 104, Théodore Monod (1902–2000) who died aged 98, and Wilfred Thesiger (1910–2003) who died aged 93.

## CONCLUDING REMARKS

Camels are remarkable animals. They look remarkable and they have many remarkable adaptations. This made them a favoured subject for painters in the Victorian era, such as Léon Belly (**Figure 4.2**). Some travellers, such as Benedict

**Figure 4.2** Léon Belly's painting of pilgrims going to Mecca, 1861.

**Figure 4.3** Camels at a watering hole near Moyale in northern Kenya. In areas such as this there are signs that climate change is leading to an increased appreciation of the value of camels in marginal drylands. *Source: Author.*

Allen in Namibia and Michael Asher and his wife in North Africa, have done heroic treks with camels in recent years.

Since their domestication a few thousands of years ago, they have proved immensely useful in a wide range of ways. They have played a pivotal role in various economies and societies in dryland areas, they have been used by explorers in northern and eastern Africa, Arabia, Australia, and in a belt of country stretching along the Silk Road from Iran to China. They have also been employed since the days of ancient Rome and some of the great Persian empires in military and policing operations. At the present time, in some parts of the world, their numbers and utilisation are decreasing, but there are other parts of the world where their numbers are growing. Some of their roles have been overtaken by, for example, the use of motor-vehicles for long-distance travels and the transport of goods, but there is an increasing realisation in other areas that the usefulness of camels still remains, not least with respect to meat and milk production. Moreover, as the world's drylands become even drier as a result of global warming, camels may become still more valuable, as they already appear to be doing in northern Kenya (**Figure 4.3**).

# REFERENCES AND READING

Abri, M.A.A. and Faye, B., 2019. Genetic improvement in dromedary camels: challenges and opportunities. *Frontiers in Genetics* 10, 167.

Acland, P.B.E., 1932. Notes on the camel in the eastern Sudan. *Sudan Notes and Records* 15, 119–149.

Afshar, A., 1978. Camels at Persepolis. *Antiquity* 52, 206–228.

Al Yahya, E., 2006. *Travellers in Arabia. British Explorers in Saudi Arabia.* Stacey International, London.

Al-Hajri, H. S., 2003. *Oman through British Eyes: British Travel Writing on Oman from 1800 to 1970* (Doctoral dissertation, University of Warwick).

Allen, M. 1972., *Palgrave of Arabia. The Life of William Gifford Palgrave 1826–88.* Macmillan, London,

Almásy, L., 1934. *The Unknown Sahara.* Franklin Tarsulat, Budapest.

Almathen, F., Charruau, P., Mohandesan, E., Mwacharo, J.M., Orozco-terWengel, P., Pitt, D., Abdussamad, A.M., Uerpmann, M., Uerpmann, H.P., De Cupere, B. and Magee, P. , 2016. Ancient and modern DNA reveal dynamics of domestication and cross-continental dispersal of the dromedary. *Proceedings of the National Academy of Sciences* 113, 6707-6712.

Andrews, R.C., 1927. Explorations in Mongolia: a review of the Central Asiatic Expeditions of the American Museum of Natural History. *Geographical Journal* 69, 1–19.

Andrews, R.C., 1932. *The New Conquest of Central Asia. A Narrative of the Explorations of the Central Asiatic Expeditions in Mongolia and China, 1921–1930.* American Museum of Natural History, New York.

Andreyev, A.I., Baskhanov, M. and Yusupova, T., 2018. *The Quest for Forbidden Lands: Nikolai Przhevalskii and his Followers on Inner Asian Tracks.* Brill, Leiden.

Anjum, N.A., 2006. Camel as transport animal in Mughal India. *Proceedings of the Indian History Congress* 67, 244-250.

Anon., 1885. Dr Gustav Nachtigal. *Proceedings of the Royal Geographical Society of London* 7, 466.

Anon., 1894. Baker, S.W. Obituary. *Nature* 49, 227–228.

Anon., 1914. Obituary: Fernand Foureau. *Geographical Journal* 43, 587–588.

Anon., 1926. Mr. C. M. Doughty. *Nature* 117, 204.

Anon., 1937. Obituary, J.F. Keane. *The Times*, November 23, 1937.

Anon., 1978. Obituary: Lord Rennell of Rodd, KBE, CB, JP. *Geographical Journal* 144, 392–393.

Anon., 1937. Obituary, Sir Percy Cox. *Geographical Journal* 90, 1–5.

Anon., Undated. *Pictorial Records of the English in Egypt*. James Sangster, London.

Atkins, W., 2018. *The Immeasurable World*. Faber and Faber, London.

Auchterlonie, P., 2011. Joseph Pitts: Exeter's First Orientalist? *Oriente Moderno* 91, 171–180.

Auchterlonie, P., 2012. *Encountering Islam: Joseph Pitts: An English Slave in 17th-Century Algiers and Mecca*. Arabian Publishing, London.

Bagnold, R.A., 1931. Journeys in the Libyan Desert 1929 and 1930. *Geographical Journal* 78(1), 13–33.

Bagnold, R.A., 1966. Obituary: G.W. Murray. *Geographical Journal* 132, 333–334.

Baker, S.W., 1866. *The Albert Nyanza Great Basin of the Nile and Exploration of Nile Sources*. Macmillan, London.

Baker, S.W., 1874. *Ismailia: A Narrative of the Expedition to Central Africa for the Suppression of the Slave Trade; Organized by Ismail, Khedive of Egypt; in 2 Volumes*. Macmillan, London.

Bald, M., 2010. *From the Sahara to Samarkand. Selected Travel Writings of Rosita Forbes, 1919–1937*. Axios Press, Mount Jackson, VA.

Ball, J., 1924. Note on the cartographical results of Hassanein Bey's journey. *Geographical Journal* 64, 367–386.

Barth, H., 1859. *Travels and Discoveries in North and Central Africa: From the Journal of an Expedition Undertaken Under the Auspices of HBM's Government, in the Years 1849–1855*. J.W. Bradley, London.

Barth, H., 1890. *Travels*. Ward Lock, London.

Bartosiewicz, L. ,2014. Camels in the front line. *Anthropozoologica* 49(2), 297-302.

Bates, O. ,1915. XXV. Ethnographic Notes from Marsa Maṭrūḥ. *Journal of the Royal Asiatic Society* 47, 717–739.

Baum, D. ,2011. The status of the camel in the United States of America. In *Proceedings of the Camel Conference*, London, UK, 23-25. https://www.soas.ac.uk/camelconference2011/file68428.pdf (accessed August 2021).

Baum, D., 2013. The camel saddle: a study. https://www.soas.ac.uk/camelconference2013/file88887.pdf (accessed August 2021).

Baum, D., 2018. Art of saddling a camel. *Aramco World* 69(6), 28–29.

Beadnell, H.J.L., 1909. *An Egyptian Oasis: An Account of the Oasis of Kharga in the Libyan Desert, With Special Reference to Its History, Physical Geography, and Water-Supply*. J. Murray, London.

Beadnell, H.J.L., 1927. *The Wilderness of Sinai*. Arnold, London.

Beard, J.S., 1970. The penetration of the Western Deserts of Australia. *Geographical Journal* 136, 557–568.

Belgrave, C.D., 1923. Siwa, the Oasis of Jupiter Ammon. John Lane, London.

Bell, G.L., 1907. *The Desert and the Sown.* William Heinemann, London.

Bell, G.L., 1911. *Amurath to Amurath.* William Heinemann, London.

Bell, M.S., 1890. The Great Central Asian trade route from Peking to Kashgaria. *Proceedings of the Royal Geographical Society* 12, 57–93.

Bellefonds, M.A.L. de, 1868. *L'Etbaye, Country Inhabited by the Bicharieh Arabs: Geography, Ethnology, Gold mines.* Arthus Bertrand, Paris.

Belzoni, G.B., 1835. *Narrative of the Operations and Recent Discoveries within the Pyramids, Temples, Tombs, and Excavations, in Egypt and Nubia: And of a Journey to the Coast of the Red Sea, in Search of the Ancient Berenkè; and Another to the Oasis of Jupiter Ammon.* H. Remy, Brussels.

Benson, L.K., 2008. *Across China's Gobi.* EastBridge, Norwalk, CT.

Bent, J.T., 1892. *The Ruined Cities of Machonaland.* Longmans, Green and Co., New York.

Bent, J.T., 1895. Exploration of the Frankincense Country, Southern Arabia. *Geographical Journal* 6, 109–133.

Bent, J.T. and Bent, T., 1900. *Southern Arabia, Soudan and Sakotra.* Smith Elder, London.

Beuk, G. ,2012. Arthur Weigall: a man out of time.

Bevis, R., 1972. Spiritual Geology: C.M. Doughty and the Land of the Arabs. *Victorian Studies,* 163–181.

Bey, A., 1816. *Travels of Ali Bey in Morocco, Tripoli, Cyprus, Egypt, Arabia, Syria, and Turkey, Between the Years 1803 and 1807.* Longman, Hurst, Rees, Orme, and Brown, London.

Blanch, L., 1954. *The Wilder Shores of Love.* Carroll and Graf, New York.

Blunt, A., 1879. *Bedouin Tribes of the Euphrates.* John Murray, London.

Blunt, A., 1881. *A Pilgrimage to Nejd, the Cradle of the Arab Race.* John Murray, London.

Blunt, W. S., 1880. A visit to Jebel Shammar (Nejd). New routes through Northern and Central Arabia. *Proceedings of the Royal Geographical Society* 2, 81–102.

Boobbyer, P., 2021. *The Life and World of Francis Rodd, Lord Rennell (1895–1978): Geography, Money and War.* Anthem Press, London.

Bourcart, J., 1951. Conrad Kilian (1898–1950). *Bulletin de la Société Géologique de France* 6(4–6), 303–312.

Boustead, J.E.H., 1934. The Camel Corps of the Sudan Defence Force. *Royal United Services Institution Journal* 79, 547-557.

Boustead, H., 1971. *The Wind of Morning.* Chatto and Windus, London.

Bovill, E.W., 1927. The Moorish invasion of the Sudan. *Journal of the Royal African Society* 26, 245–262.

Bovill, E.W. , 1956. The camel and the Garamantes. *Antiquity* 30, 19-21.

Braun, K. and Passon, J. (eds.), 2020. *Across the Sahara. Tracks, Trade and Cross-Cultural Exchange in Libya.* Springer, Cham, Switzerland.

Brent, P., 1978. *Far Arabia. Explorers of the Myth.* Readers Union, Newton Abbot.

Brodie, F. M., 1967. *The Devil Drives.* Eyre & Spottiswoode, London.

Brogan, O., 1954, The camel in Roman Tripolitania. *Papers of the British School at Rome* 22, 126-131

Brower, D., 1994. Imperial Russia and its orient: the renown of Nikolai Przhevalsky. *Russian Review* 53, 367–381.

Browne, W.G., 1799. *Travels in Africa, Egypt and Syria, from the Years 1792 to 1798.* Cadell and Davies, London.

Bruce, J., 1790. *Travels to Discover the Source of the Nile in the Years 1768, 1769, 1770, 1771, 1772, and 1773.* Volume 4. Robinson, Edinburgh and London.

Bruce-Lockhart, J. and Wright, J., 2000. *Difficult and Dangerous Roads.* Sickle Moon Books, London.

Buchanan A., 1921. *Exploration of Aïr: Out of the World North of Nigeria.* John Murray, London.

Buchanan, A., 1926. *Sahara.* John Murray, London.

Buffon, G-L., 1797. *Histoire Naturelle.* Translated into English in ten volumes. Barr, London.

Bulliet, R.W., 1990. *The Camel and the Wheel.* Columbia University Press, New York.

Burckhardt, J.L., 1819. *Travels in Nubia.* John Murray, London.

Burckhardt, J.L., 1822. *Travels in Syria and the Holy Land.* John Murray, London.

Burckhardt, J.L., 1829. *Travels in Arabia.* John Murray, London.

Burger, P.A., 2016. The history of Old World camelids in the light of molecular genetics. *Tropical Animal Health and Production* 48, 905–913.

Burger, P.F., 2017. 'Tavernier, Jean-Baptiste'. *Encyclopædia Iranica*, online edition, 2017. Available at http://www.iranicaonline.org/articles/tavernier-jean-baptiste.

Burke and Wills Web. http://www.burkeandwills.net.au/index.php (accessed August 2021).

Burton, R.F., 1854. Journey to Medina, with route from Yambu. *Journal of the Royal Geographical Society of London* 24, 208–225.

Burton, R.F., 1855. *Personal Narrative of a Pilgrimage to Al-Madinah and Mecca.* Longman Green, London.

Burton, R.F., 1874. *Personal Narrative of a Pilgrimage to Mecca and Medina.* B. Tauchnitz, Leipzig.

Buxton, E.N., 1894. In the mountains of Egypt. *The Nineteenth Century and After: A Monthly Review* 35, 495–512.

Cable, M., 1942. A new era in the Gobi. *Geographical Journal* 100, 193–205.

Cable, M. and French, E., 1942. *The Gobi Desert.* Hodder and Stoughton, London.

Cailliaud, F., 1822. *Travels in the Oasis of Thebes, and in the Deserts Situated East and West of the Thebaid: In the Years 1815, 16, 17, and 18.* R. Phillips & Company, London.

Caillié, R., 1830. *Travels through Central Africa to Timbuctoo; and across the Great Desert to Morocco; performed in the years 1824–1828.* 2 volumes. Colburn and Bentley, London.

Çaliskan, V., 2009. Geography of a Hidden Cultural Heritage: Camel Wrestles in Western Anatolia. *Journal of International Social Research* 2, 123–137.

Canton, J., 2011. *From Cairo to Baghdad. British Travellers in Arabia.* I.B. Tauris, London.

Capenny, S.H.F., 1902. The Nushki-Sistan route from Quetta to eastern Persia. *Scottish Geographical Magazine* 18, 208-217.

Cardona, M.K., 2012. Alexine Tinné: Nineteenth-Century explorer of Africa. *Terrae Incognitae* 44(2), 124–138.

Carman, B. and McPherson, J. (eds), 1983. *Bimbashi McPherson.* BBC, London.

Carnegie, D.W., 1898a. Explorations in the interior of Western Australia, between Coolgardie and Kimberley. *Scottish Geographical Magazine* 14, 113–136.

Carnegie, D.W., 1898b. *Spinifex and Sand.* C. Arthur Pearson, London.

Carroll, C.C., 1903. *The Government's Importation of Camels: a Historical Sketch.* US Government Printing Office, Washington, D.C.

Carruthers, D., 1910. A journey in north-western Arabia. *Geographical Journal* 35, 225–245.

Carruthers, D., 1922. Captain Shakespear's last journey. *Geographical Journal* 59, 321–334, 401–418.

Carruthers, D., 1958. Reminiscences of Gertrude Bell. *Journal of the Royal Central Asian Society* 45, 52–57.

Cavendish, H.S.H., 1898. Through Somaliland and around and south of Lake Rudolf. *Geographical Journal* 11, 372–393.

Chamberlain, P.G., 2018. Carsten Niebuhr and the Danish Expedition to Arabia. *Aramco World* 69 (1), 18–25.

Christie, T.L., 1967. Shaikh Burckhardt: explorer. *Saudi Aramco World*, 18 (5), 15–17.

Chudeau, R., 1919. La capture du Niger par le Taffassasset. *Annales de Géographie* 28, 52–60

Chudeau, R., 1920. Etude sur les dunes sahariennes. *Annales de Géographie* 29, 334–351

Chudeau, R., 1921. Climate changes in the Sahara during the Quaternary. *CR Acad. Sci.* 172, 604–607.

Clerk, G., 1944. Address at the Annual General Meeting of the Society, Held on 19 June 1944. *Geographical Journal* 104, 1–7.

Clifton Pierce, G., 1901. *Field Genealogy*, vol II. Hammon Press, Chicago.

Connelly, T.L., 1966. The American camel experiment: a reappraisal. *The Southwestern Historical Quarterly* 69, 442–462.

Cox, P., 1925. Some excursions in Oman. *Geographical Journal* 66, 193–221.

Craig, J., 2008. Philby, Harry St John Bridger (1885–1960). *Oxford Dictionary of National Biography Online.*

Crisp, N.J., 2017. *Out of Oscurity: John F.T. Kean's Late Nineteeth Century Hajj.* Masters thesis, John Washington University.

Dabrowa, E., 1991. Dromedarii in the Roman Army: A Note. In *Roman Frontier Studies 1989: Proceedings of the XV*[th] *International Congress of Roman Frontier Studies* (p. 364). University of Exeter Press, Exeter.

Dalrymple, W., 2013. *Return of a King.* Bloomsbury, London.

Darlow, M. and Bray, B., 2010. *Ibn Saud. The Desert Warrior and his Legacy.* Quartet Books, London.

Davidson, K., 2000. Huntington, Ellsworth. *American National Biography on line.*

Davies, H.R.J., 2006. W.G. Browne's visit to Darfur 1793–1796. *Sudan Studies* 34, 2–13.

De Villers, M. and Hirtle, S., 2003. *Sahara. The Life of the Great Desert.* Harper Collins, London.

Deasey, D., 1976. Warburton, Peter Egerton (1813–1889). In *Australian Dictionary of Biography*, Volume 6. National Centre of Biography, Australian National University.

Degen, A.A., El-Meccawi, S. and Kam, M., 2019. The Changing Role of Camels among the Bedouin of the Negev. *Human Ecology* 47, 193–204.

Della Cella, P., 1822. *Narrative of an Expedition from Tripoli in Barbary, to the Western Frontier of Egypt, in 1817: By the Bey of Tripoli; in Letters to Dr. Viviani of Genoa.* John and Arthur Arch, London.

Denham, D., 1828. *Narrative of Travels and Discoveries in Northern and Central Africa in the Years 1822, 1823 and 1824.* 3rd Edition. John Murray, London.

Desio, A., 1950. *Le Vie Della Sete.* Edizione Hoepli, Milan.

Devonshire Association 1920 – biography by Cicely Radford and bibliography. http://archive.org/stream/ReportTransactionsOfTheDevonshireAssociationVol521920/TDA1920vol52#page/n247/mode/2up (accessed August 2021).

Dickson, W.K., 1901. *Life of Major-General Sir Robert Murdoch Smith, KEMG, Royal Engineers.* Blackwood.

Dioli, M., 2018. Camels in Europe: A brief history of camel hybrids and camel breeding in Europe over the last 1000 years. In *Second Selçuk-Ephesus Symposium on culture of camel-dealing and camel wrestling, 18–20 January 2018, Selçuk, Izmir*, 8–11. Available online.

Doughty, C.M., 1884. Travels in north-western Arabia and Nejd. *Proceedings of the Royal Geographical Society* 6, 382–399.

Doughty, C.M., 1888. *Arabia Deserta.* Cambridge University Press, Cambridge.

Douls, C., 1888. *Voyages dans le Sahara Occidental et le Sud Marocian.* Cagniard, Rouen.

Duveyrier, H., 1864. *Exploration du Sahara: les Touareg du Nord.* Challamel, Paris.

Eccles, G.J., 1927. The Sultanate of Muscat and 'Oman: With a description of a journey into the interior undertaken in 1925. *Journal of the Central Asian Society* 14, 19–42.

Edwards, G.P., Zeng, B., Saalfeld, W.K. and Vaarzon-Morel, P., 2010. Evaluation of the impacts of feral camels. *Rangeland Journal* 32, 43–54.

Eriksen, M., Lusher, A., Nixon, M. and Wernery, U., 2021. The plight of camels eating plastic waste. *Journal of Arid Environments* 185, February, 104374.

Fabens, J.W., 1865. *The Uses of the Camel: Considered with a View to His Introduction Into Our Western States and Territories.* Carleton, New York.

Faye, B. 2014., The camel today: assets and potentials. *Anthropozoologica* 49, 167–176.

Faye, B. 2015., Role, distribution and perspective of camel breeding in the third millennium economies. *Emirates Journal of Food and Agriculture* 27, 318–327.

Faye, B., 2020. How many large camelids in the world? A synthetic analysis of the world camel demographic changes. *Pastoralism* 10, 1–20.

Field, H., 1876. From the Lakes of Killarney to the Golden Horn. Scribner, Armstrong, New York. *Field Magazine*: https://www.thefield.co.uk/country-house/the-imperial-camel-corps-31175 (accessed August 2021).

Field, H., 1877. *From Egypt to Japan.* Scribner Armstrong, New York.

Field, H., 1888. *On the Desert – Recent Events in Egypt.* Scribner Armstrong, New York.

Fiennes, R., 2015. *Heat.* Simon & Schuster, London.

Fitzpatrick, K., 1969. Burke, Robert O'Hara (1821–1861). *Australian Dictionary of Biography*, Volume 3, Melbourne University Press, Melbourne, 301–303.

Flannery, T., 1998. *The Explorers. Stories of Discovery and Adventure from the Australian Frontier.* Grove Press, New York.

Fleming, F., 2003. *The Sword and the Cross.* Grants Books, London.

Fookes, P., Waring, D. and Lee, M., 2019. Wilfred Thesiger and the Umm as Samim, Oman. *Geology Today* 35, 221–227.

Forbes, R., 1921. *The Secret of the Sahara: Kufara.* Cassell, London.

Förster, F. and Riemer, H. (eds.), 2013. *Desert Road Archaeology in Ancient Egypt and Beyond.* Heinrich-Barth-Institut, Cologne.

Forth, N. de R., 1991. *A Fighting Colonel of Camel Corps.* Merlin Books, Braunton, Devon.

Forth, N.D.L., 1930. More Journeys in Search of Zerzura. *Geographical Journal* 75, 48–59.

Foureau, F., 1894. *Rapport Sur La Mission Au Sahara et Chez Les Touareg Azdjer: 7 October 1893–Mars 1894.* Challamel, Paris.

Foureau, F., 1897. *Au Sahara: Mes Deux Missions de 1892 et 1893: le Gassi Touil et le Grand Erg: l'Oudje sud et la Tinghert: Hassi Messegguem et Hassi Imoulay: Réédition du Rapport de Mission Publié en Juillet 1893.* Challamel, Paris.

Foureau, F., 1902. *D'Alger au Congo par le Tchad.* Masson, Paris.

Foureau, F., 1905. *Documents Scientifiques de la Mission Saharienne: Mission Foureau-Lamy.* Masson, Paris.

Freeth, Z. and Winstone, V., 1978. *Explorers of Arabia.* Allen and Unwin, London.

French, P., 1994. Younghusband. *The Last Great Imperial Adventurer.* Harper Collins, London.

Furlong, C.W., 1909. *The Gateway to the Sahara, Adventures and Observations in Tripoli.* Scribners, New York.

Garba, M.M., 2020. The role of Hanns Vischer (Dan Hausa) on the development of secular education in the Protectorate of Northern Nigeria. *International Journal of Research and Scientific Innovation* 7(1), 212–215.

Gardner, B., 1968. *The Quest for Timbuctoo.* Cassell, London.

Garnett, R., 2004a. Browne, William George (1768–1813). *Oxford Dictionary of National Biography Online.*

Garnett, R., 2004b. Burke, Robert O'Hara (1820–1861). *Oxford Dictionary of National Biography Online.*

Gauthier-Piltrers H. and Dagg. A.I., 1981. *The Camel.* University of Chicago Press, Chicago and London.

Gautier E.-F., 1921. René Chudeau (1864–1921). *Annales de Géographie* 30, 390–391.

Gautier, E.-F. and Chudeau, R., 1908. *Missions in the Sahara. 1. Algerian Sahara.* Armand Colin, Paris.

Gearon, E., 2011. *The Sahara. A Cultural History.* Signal Books, Oxford.

Ghose, I., 2006. Imperial Player: Richard Burton in Sindh. In T. Youngs (ed) *Travel Writing in the Nineteenth Century: Filling the Blank Spaces,* Chapter 5. Anthem, London.

Giles, E., 1889. *Australia Twice Traversed,* 2 vols. Sampson Low, Marston, Searle and Rivington, London.

Gladstone, P., 1970. *Travels of Alexine. Alexine Tinne, 1835–1869.* John Murray, London.

Gleichen, L.E., 1888. *With the Camel Corps up the Nile.* Chapman & Hall, London.

Godsall, J. R., 1993. Fact and fiction in Richard Burton's Personal Narrative of a Pilgrimage to El-Medinah and Meccah (1855–6). *Journal of the Royal Asiatic Society* (Third Series) 3, 331–351.

Goldberg, J., 1986. Captain Shakespear and Ibn Saud: a balanced reappraisal. *Middle Eastern Studies* 22, 74–88.

Goldsmit, F.J., 1892. Sir Lewis Pelly. *Proceedings of the Royal Geographical Society* 14, 416–421.

Gondwana History, 2011. Camel expedition conquers highest dunes on earth. *Gondwana History,* Gondwana Publishers, Windhoek, 43–46.

Gosse, W.C., 1874. *Report and Diary of Mr. WC Gosse's Central and Western Exploring Expedition, 1873.* Government Printer, Adelaide.

Goudie, A., 2008. *Wheels across the Desert. Exploration of the Libyan Desert by Motor Car.* Silphium Press, London.

Goudie, A., 2016. *Great Desert Explorers*. Silphium Press and Royal Geographical Society, London.

Grant, J.A., 1884. Route March, with camels, from Berber to Korosko in 1863. *Proceedings of the Royal Geographical Society and Monthly Record of Geography* 6, 326–335.

Grant, W.O., Harrison, J.J., 1901. On the birds collected by Mr. J.J. Harrison between Zeila and Lakes Rudolf and Baringo, Eastern Africa. *Ibis* 43, 278–99

Gray, A.A., Farquhar, F.P. and Lewis, W.S., 1930. *Camels in Western America*. California Historical Society, San Francisco.

Green, G., 2004. Harrison, James Jonathan. *Oxford Dictionary of National Biography*. https://doi.org/10.1093/ref:odnb/57208.

Green, G.W., 1885. The organization and employment of camel corps in warfare. *Royal United Services Institution Journal* 29, 521–537.

Grigson, C., 1983. A very large camel from the Upper Pleistocene of the Negev Desert. *Journal of Archaeological Science* 10, 311–316.

Grove, A.T. and Goudie A.S., 1971. Secrets of Lake Stephanie's past. *Geographical Magazine* 43, 542–547.

Guliye, A.Y., Noor, I.M., Bebe, B.O. and Kosgey, I.S., 2007. Role of camels (*Camelus dromedarius*) in the traditional lifestyle of Somali pastoralists in northern Kenya. *Outlook on Agriculture* 36, 29–34.

Habinger, S.G., De Cupere, B., Dövener, F., Pucher, E. and Bocherens, H., 2020. Mobility and origin of camels in the Roman Empire through serial stable carbon and oxygen isotope variations in tooth enamel. *Quaternary International* 557, 80–91.

Hall, D.N., 2001. Théodore Monod, 1902-2000. *Geographical Journal* 167, 191–192.

Hall, R., 1981. *Lovers on the Nile*. Quartet Books, London.

Hamilton, A., 2010. *An Arabian Utopia: the Western Discovery of Oman*. Oxford University Press and Arcadian Library, Oxford.

Hankey, J., 2001. *A Passion for Egypt*. I.B. Tauris, London.

Hansen, T., 1964. *Arabia Felix: The Danish Expedition of 1761–1767*. Readers Union Collins, London.

Harding, W.H., 1913. The Farafra Depression and Bu Mungar Hattia. *Geographical Journal* 42, 455–461.

Hare, J., 1998. *The Lost Camels of Tartary*. Abacus, London.

Hare, J., 2003. *Shadows across the Sahara*. Constable, London.

Harris, C.A., 2004. Warburton, Peter Egerton (1813–1889). *Oxford Dictionary of National Biography Online*.

Harrison, J.J., 1901. A journey from Zeila to Lake Rudolf. *Geographical Journal* 18, 258–275.

Harrison, J.V., 1941. Coastal Makran. *Geographical Journal* 97, 1–15.

Hart, L., 1934. 'T.E. Lawrence' in Arabia and After. Jonathan Cape, London.

Hartwig, G., 1886. *The Tropical World*. Longmans, Green and Co., London.

Harvey, D., 1983. Owen Lattimore. A memoire. *Antipode*, 15(3), 3–11.

Hassanein Bey, A.M., 1924a. Crossing the untraversed Libyan Desert. *National Geographic Magazine* 46(3), 49pp.

Hassanein Bey, A.M., 1924b. Through Kufra to Darfur. *Geographical Journal* 64, 273–291.

Hassanein Bey, A.M., 1925. *The Lost Oases*. Thomas Butterworth, London.

Heffernan, M., 1989. The limits of Utopia: Henri Duveyrier and the exploration of the Sahara in the nineteenth century. *Geographical Journal* 155, 342–352.

Hibbert, C., 1982. *Africa Explored. Europeans in the Dark Continent, 1769–1889.* Hamish Hamilton, London.

Hill, J.G., 1905. *The Calvert Scientific Exploring Expedition (Australia, 1896).* G. Philip & Son, Limited, London.

Hogarth, D.G., 1904. *The Penetration of Arabia. A Record of the Development of Western Knowledge Concerning the Arabian peninsula.* Stokes and Company, New York.

Hogarth, D.G., 1926a. Obituary. Gertrude Lowthian Bell. *Geographical Journal* 68, 363–368.

Hogarth, D.G., 1926b. Obituary. Lieut.-Colonel G.E. Leachman, C.I.E., D.S.O. *Geographical Journal* 56, 325.

Hogarth, D.G., 1927. Gertrude Bell's journey to Hayil. *Geographical Journal* 70, 1–17.

Holdich, T. H., 1909. What we have learnt from Dr Sven Hedin. *Geographical Journal* 33, 436–438.

Hopkinson, H.C.B., 1899. The Sirdar's Camel Corps. *The Cornhill Magazine* 6, 433–448.

Hornemann, F., 1802. *Travels from Cairo to Mourzouk.* African Association, London.

Hough, W., 1841. *A Narrative of the March and Operations of the Army of the Indus, in the Expedition to Affghanistan in the Years 1838–1839: Comprising Also the History of the Dooranee Empire from Its Foundation to the Present Time.* Wm. H. Allen, London

Howell, G., 2006. *Daughter of the Desert.* Macmillan, London.

Howell, G., 2015. *Gertrude Bell. A Woman in Arabia. The Writings of the Queen of the Desert.* Penguin Books, New York.

Hureau, J.C., 2001. Un exceptionnel naturaliste éclectique. *Autres Temps. Cahiers d'éthique Sociale et Politique* 70, 25–38.

İnal, O. 2020. One-Humped history: The camel as historical actor in the Late Ottoman Empire. *International Journal of Middle East Studies*, 1–16.

Inchbald, G., 2005. *With the Imperial Camel Corps in the Great War.* Leonaur, London.

Irwin, R., 2010. *Camel.* Reaktion Books, London.

Jarry, I., 2001. *Théodore Monod.* Payot, Paris.

Jarvis, C.S., 1938. *Desert and Delta.* John Murray, London.

Kagunyu, A.W. and Wanjohi, J., 2014. Camel rearing replacing cattle production among the Borana community in Isiolo County of Northern Kenya, as climate variability bites. *Pastoralism* 4, 1–5.

Keane J.F., 1881a. *Six Months in Meccah: An Account of the Mohammedan Pilgrimage to Meccah. Recently Accomplished by an Englishman Professing Mohammedanism.* Tinsley Brothers, London.

Keane, J.F., 1881b. *My Journey to Medinah.* Tinsley Brothers, London.

Kearton, C., 1924. *The Shifting Sands of Algeria.* Arrowsmith, London.

Keay, J. 1982., *Eccentric Travellers.* John Murray, London.

Keays, G.A.V., 1939. Note on the history of the camel corps. *Sudan Notes and Records* 22, 103-123.

Kelly, J.B., 1977. Review of Philby of Arabia by Elizabeth Monroe. *Middle Eastern Studies* 13, 144-146.

Kennedy, D., 2013. *The Last Blank Spaces.* Harvard University Press, Cambridge and London.

Kiernan, R.H., 1937. *The Unveiling of Arabia.* Harrap, London.
Kilian, C., 1935. Explorations Sahariennes. *Geographical Journal* 86, 1, 17-25.
King, W.H., 1902. A Visit to the Hoggar Tuaregs. *Geographical Journal* 20, 507-517.
King, W.H., 1903. *A Search for the Masked Tawareks.* Smith, Elder, & Company, London.
King, W.H., 1912. Travels in the Libyan Desert. *Geographical Journal* 39, 133-137.
King, W.H., 1913. The Libyan desert from native information. *Geographical Journal* 42, 277–283.
King, W.H., 1918. Study of a dune belt. *Geographical Journal* 51, 16-33.
King, W.H., 1925. *Mysteries of the Libyan Desert.* London, Seeley, Service and Co.
King, W.H., 1928. Lost oases of the Libyan Desert. *Geographical Journal* 72, 244-249.
King, W.H., 1930. The lost oases. *Geographical Journal* 75, 61-64.
King, W.H., 1931. Pioneer desert exploration. *Geographical Journal* 77, 541-547.
King, W.H., 1933. Obituary. Lieut.-Col. Nowell Barnnard de Lancey Forth. *Geographical Journal* 81, 479-480.
Köhler-Rollefson, I., 1992. The Raika dromedary breeders of Rajasthan: A pastoral system in crisis. *Nomadic Peoples* 30, 74-83.
Köhler-Rollefson, I., 2018. The Raikas and camels in Rajasthan. https://www.sahapedia.org/camel-cultures-of-india (accessed February 2021).
Lander, R. 1830. *Records of Captain Clapperton's Last Expedition to Africa: By Richard Lander, His Faithful Attendant, and the Only Surviving Member of the Expedition: With the Subsequent Adventures of the Author.* Henry Colburn and Richard Bentley, London.
Larner, A.J. and Fisher, H.J., 2000. Dr Gustav Nachtigal (1834–1885): A contribution to the history of medicine in mid-nineteenth-century Africa. *Journal of Medical Biography* 8, 43–48.
Laughton, J.K., 2004. Wellsted, James Raymond (1805-1842). *Oxford Dictionary of National Biography Online.*
Lawrence, T.E., 1919. Notes on camel journeys. *Arab Bulletin* No 111, 71-2.
Lawrence, T.E., 1927. *Revolt in the Desert.* Jonathan Cape, London.
Lawrence, T.E., 1935. *Seven Pillars of Wisdom.* Jonathan Cape, London.
Leese, A.S., 1927 *A Treatise on the One-Humped Camel in Health and in Disease.* Haynes, Stamford.
Lelubre, M., 1992. Conrad Kilian, géologue et explorateur saharien. *Travaux du Comité français d'Histoire de la Géologie*, 3ème série (tome 6), 75–85.
Lenz, O., 1886. *Timbouctou: Voyage au Maroc, au Sahara, et au Soudan*, traduit de l'Allemand . . . par Pierre Lehautcourt. 2 vols. Librairie Hachette, Paris.
Leonard, A.G., 1894. *The Camel, its Uses and Management.* Longmans, Green and Company, London.
Lévy-Jahanbakht, D., 2017. Expéditions scientifiques dans les déserts iraniens et le mont Elbourz entre 1925 et 1939. *Recherches Germaniques* 12, 141–152.
Lewis, W.S., 1928. The camel pack trains in the mining camps of the west. *Washington Historical Quarterly* 19, 271–284.
Lewis, W.S., Gray, A.A. and Farquhar, F.P., 1930. A contribution towards a bibliography of the camel. *California Historical Society Quarterly* 9, 336–344.
Longford, E., 2007. *A Pilgrimage of Passion: the Life of Wilfrid Scawen Blunt.* I.B. Tauris, London.
Lukitz, L., 2004. Bell, Gertrude Margaret Lowthian (1868–1926). *Oxford Dictionary of National Biography Online.*

Lyon, G.F., 1821. *Narrative of Travels in Northern Africa, in the Years 1818, 19, and 20.* John Murray, London.

Lyon, M.J., 2017. *Like a Rolling Stone. The Life and Travels of George Francis Lyon, RN, FRS.* Publisher unknown.

Marković, N., Ivanišević, V., Baron, H., Lawless, C. and Buckley, M., 2021. The last caravans in antiquity: Camel remains from Caričin Grad (Justiniana Prima). *Journal of Archaeological Science: Reports* 38, 103038.

Marozzi, J., 2010. Heinrich Barth. Crossing the Sahara (1821–1865). In R. Hanbury-Tenison (ed). *The Great Explorers.* Thames and Hudson, London, 196–203.

Marsh, G.P., 1855. The Camel. *9th Annual Report of the Smithsonian Institution*, 99–122.

Marsh, G.P., 1856. *The Camel: His Organization, Habits and Uses Considered with Reference to His Introduction Into the United States.* Gould and Lincoln, Boston.

Marsh, G.P., 1864. *Man and Nature.* Scribner, New York.

Martin, G.J., 1973. *Ellsworth Huntington: His Life and Thought.* Archon Books, Hamden, CT.

Martini, P., 2019. Camel fossils from the El Kowm Basin, Syria. Diversity and evolution. Thesis, University of Basel.

Mason, M.H., 1936. *The Paradise of Fools. Being an Account of the Expedition which Covered 6,300 Miles of the Libyan Desert by Motor-car in 1935.* Hodder & Stoughton, London.

Massey, W.T., 1918. *The Desert Campaigns.* Constable, London.

Mathuisieulx, M. de, 1904. An expedition to Tripoli. *Bulletin of the American Geographical Society* 36(12), 736–744.

Mattingly, D.J. (ed.), 2013. *The Archaeology of Fazzān*, volume 4. Department of Antiquities, Tripoli and Society for Libyan Studies, London.

Maud, P., 1904. Exploration in the southern borderland of Abyssinia. *Geographical Journal* 23, 552–579.

McGuirk, R., 2007. *The Sanussi's Little War.* Arabian Publishing, London.

McKnight, T.L., 1969. *The Camel in Australia.* Melbourne University Press, Carlton, Victoria.

McLynn, F., 1990. *Burton: Snow upon the Desert.* John Murray, London.

Meissas, G., 1894. *Les Grands Voyages Contemporains.* Hachette, Paris.

Meyer, K.E. and Brysac, S.B., 1999. *Tournament of Shadows.* Counterpoint, Washington, D.C.

Miles, S.B., 1874. Journey from Gwadur to Karachi. *Journal of the Royal Geographical Society of London* 44, 163–182.

Miles, S.B., 1896. Journal of an excursion in Oman, in South-East Arabia. *Geographical Journal* 7, 522–537.

Miles, S.B., 1910. On the border of the great desert: a journey in Oman. *Geographical Journal* 36, 159–178.

Miles, S.B., 1919. *The Countries and Tribes of the Persian Gulf.* Harrison and Sons, London.

Miles, S.B. and Munzinger, M.W., 1871. Account of an excursion into the interior of Southern Arabia. *Journal of the Royal Geographical Society of London* 41, 210–245.

Monod, T., 1928. Une traversée de la Mauritanie occidentale. *Revue de Géographie Physique et de Géologie Dynamique* 1, 3–25, 88–106.

Monod, T., 1932. *L'Adrar Ahnet – Contribution à l'étude archéologique d'un District Saharien*, Institut d'Ethnologie, Paris.

Monod, T., 1937. *Méharées – Explorations au vrai Sahara.* Paris.

Monod, T., 1958, *Majabat Al-Koubra: Contribution a l'Etude de l'Empty Quarter Ouest-Saharien.* IFAN, Dakar.

Monod, T., 1984. *L'Emeraude des Garamantes. Souvenirs d'un Saharien.* Harmattan, Paris.

Monod, T., 1997. *Terre et Ciel.* Actes Sud, Arles.

Monod, T., and Diemer, E., 2000. *Zerzura: L'Oasis Légendaire du Désert Libyque.* Paris: Editions Vent de Sable.

Monroe, E., 1973a. Across the Rub 'Al Khali. Saudi *Aramco World* 24, 6–13.

Monroe, E., 1973b. *Philby of Arabia.* Faber and Faber, London.

Morrison, A., 2014. Camels and colonial armies. The Logistics of warfare in Central Asia in the early 19th century. *Journal of the Economic & Social History of the Orient* 57, 443–485.

Mukasa-Mugerwa, E., 1981. *The Camel (*Camelus dromedarius*): a Bibliographical Review.* International Livestock Centre for Africa, Addis Ababa.

Murray, G.W., 1925. The Roman roads and stations in the Eastern Desert of Egypt. *The Journal of Egyptian Archaeology* 11, 1, 138–150.

Murray, G.W., 1926a. Aidhab. *Geographical Journal* 68, 235-240.

Murray, G.W., 1926b. Graves of oxen in the Eastern Desert of Egypt. *The Journal of Egyptian Archaeology* 12, 248–249.

Murray, G.W., 1931. A small temple in the Western Desert. *The Journal of Egyptian Archaeology* 17, 81–84.

Murray, G.W., 1935. *Sons of Ishmael: A Study of the Egyptian Bedouin.* Routledge, London.

Murray, G.W., 1939. The road to Chephren's quarries. *Geographical Journal* 94, 97–111.

Murray, G.W. (ed.), 1950. *The Survey of Egypt, 1898-1948.* Survey Dept. Paper No. 50, Cairo.

Murray, G.W., 1960. Obituary: Wilfrid Jennings Bramly, M.B.E., M.C. *Geographical Journal* 126, 258–259.

Murray, G.W., 1967. *Dare Me to the Desert.* George Allen and Unwin, London.

Murray, G.W. and Derry, D.E., 1923. A Pre-Dynastic Burial on the Red Sea Coast of Egypt. *Man*, 23, 129–131.

Murray, G.W. and Warmington, E.H., 1967. Trogodytica: the Red Sea littoral in Ptolemaic times. *Geographical Journal* 133, 24–33.

Nachtigal, G., 1876. Journey to Lake Chad and neighbouring regions. *Journal of the Royal Geographical Society of London* 46, 396–411.

Nachtigal, G., 1879–1881. *Sahara und Sudan.* 3 volumes. Brockhaus, Berlin and Leipzig.

Nachtigal, G., 1974. *Sahara and Sudan*, translated by A.G.B. Fisher and H.J. Fisher. Barnes and Noble/Harper Row, New York.

Nesbitt, L.M., 1930a. Danakil traversed from South to North in 1928. *Geographical Journal* 76, 298–315, 319–414, 545–557.

Nesbitt, L.M., 1930b. *La Dancalia Explorata.* Bemporad, Florence.

Nesbitt, L.M., 1934. *Hell-hole of Creation. The Exploration of Abyssinian Danakil.* Knopf, New York.

Newbold, D., 1924a. A desert odyssey of a thousand miles. *Sudan Notes and Records* 7, 43–92.

Newbold, D., 1924b. Two journeys to Kufra Oasis. *Sudan Notes and Records* 7, 103–107.

Newbold, D. and Shaw, W.B.K., 1928. An exploration in the south Libyan Desert. *Sudan Notes and Records* 11, 103–194.

Niebuhr, C., 1774. *Reisebeschreibung nach Arabien und andern umliegenden Ländern.* Möller, Copenhagen and Hamburg.

Niebuhr, C., 1792. *Travels through Arabia and Other Countries of the East*, translated by R. Heron, London. R. Morison, Edinburgh.

Ondaatje, C., 1996. *Sindh Revisited. A Journey in the Footsteps of Captain Sir Richard Francis Burton*. Harper Collins, London.

Orlando, L., 2016. Back to the roots and routes of dromedary domestication. *Proceedings of the National Academy of Sciences* 113, 6588–6590.

Orombelli, G., 2004. Ardito Desio (1897–2001) and his contribution to physical geography and geomorphology. *Geografia Fisica e Dinamica Quaternaria* 27, 89–96.

Pacho, M.J.R., 1827. *Relation d'un Voyage dans la Marmarique, la Cyrenaique, et Oasis d'Audjelah et de Maradeh*. Firmin Didot, Paris.

Palgrave, W.G., 1864. Observations made in Central, Eastern, and Southern Arabia during a journey through that country in 1862 and 1863. *Journal of the Royal Geographical Society* 34, 111–154.

Palgrave, W.G., 1865. *Personal Narrative of a Year's Journey through Central and Eastern Arabia (1862–63)*. 2 volumes. Macmillan, London.

Parkinson, F.B., 1898. Two recent journeys in northern Somaliland. *Geographical Journal* 11, 15–48.

Parkyns, M., 1850. The Kubbabish Arabs between Dongola and Kordofan. *Journal of the Royal Geographical Society of London* 20, 254–275.

Pease, A.E., 1898. Some account of Somaliland: With notes on journeys through the Gadabürsi and western Ogaden countries, 1896–1897. *Scottish Geographical Magazine* 14, 57–73.

Peel, W., 1852. *A Ride through the Nubian Desert*. Longman, Brown, Green, and Longmans, London.

Pelly, L., 1865. A visit to the Wahabee capital, Central Arabia. *The Journal of the Royal Geographical Society of London* 35, 169–191.

Pelly, L., 1978. *Report on a Journey to Riyadh in Central Arabia*. With a new introduction by R. L. Bidwell, Oleander Press, Cambridge.

Perdue, P.C., 2018. Owen Lattimore: World Historian. *Oxford Handbooks Online*, DOI: 10.1093/oxfordhb/9780199935369.013.2.

Perham, M. and Simmons, J., 1942. *African Discovery*. Penguin Books, Harmondsworth.

Philby, H. St J.B., 1920. Southern Najd. *Geographical Journal* 55, 161–185.

Philby, H. St J.B., 1922. *Heart of Arabia*. Constable, London.

Philby, H. St J.B., 1923. Jauf and the North Arabian Desert. *Geographical Journal* 62, 241–249.

Philby, H. St J.B., 1933. *The Empty Quarter*. Constable, London.

Philby, H. St J.B., 1946. *A Pilgrim in Arabia*. Robert Hale, London.

Pigière, F. and Henrotay, D., 2012. Camels in the northern provinces of the Roman Empire. *Journal of Archaeological Science* 39(5), 1531–1539.

Pitts, J., 1704. *A True and Faithful Account of the Religion and Manners of the Mahometans*. Taylor, London.

Poncet, C.-J., 1949. *A Voyage to Ethiopia in the Years 1698, 1699 and 1700*. Hakluyt Society, London.

Porch, D., 1984. *The Conquest of the Sahara*. Jonathan Cape, London.

Porch, D., 1985. The dark side of Beau Geste – The conquest of the Sahara. *Proceedings of the Meeting of the French Colonial Historical Society* 10, 219–230.

Potts, D.T., 2004. Camel hybridization and the role of Camelus bactrianus in the ancient Near East. *Journal of the Economic and Social History of the Orient* 47, 143–165.

Prescott, H.F.M., 1958. *Once to Sinai; the Further Pilgrimage of Friar Felix Fabri.* Macmillan, New York.

Prothero, R.M. 1958. Heinrich Barth and the Western Sudan. Geographical Journal 124, 326–337.

Raunkiaer, B., 1969. *Though Wahhabiland on Camelback*, edited by G. de Gaury. Praeger, London.

Ravenstein, E.G., 1896. Gerhard Rohlfs. *Geographical Journal* 8, 184–185.

Rayfield, D., 1976. *The Dream of Lhasa. The Life of Nikolay Przhevalsky (1839–1888) Explorer of Central Asia*. Paul Elek, London.

Reid, F., 1934. *The Fighting Cameliers.* Leonaur, London.

Rennell, J., 1791. On the rate of travelling, as performed by camels; and its applications, as a scale, to the purposes of geography. *Philosophical Transactions of the Royal Society of London* 81, 129–145.

Robertson, M.J., 2014. *With the Cameliers in Palestine.* Pickle Partners Publishing.

Robinson, A.E., 1931. The 'Utfa' or camel-litter of the Arabs. *Journal of the Royal African Society* 30(118), 69–78.

Rodd, F., 1923. A Journey in Aïr. *Geographical Journal* 62, 81–101.

Rodd, F., 1926. *People of the Veil: Being an Account of the Habits, Organisation and History of the Wandering Tuareg Tribes which Inhabit the Mountains of Air, or Asben in the Central Sahara.* Macmillan, London.

Rodd, F., 1929. A second journey among the Southern Tuareg. *Geographical Journal* 73(1), 1–18.

Rohlfs, F.G., 1874. *Adventures in Morocco and Journey along the Oases of Draa and Tafilet.* Sampson Low, London.

Rohlfs, F.G., 1875. *Drei Monate in der Libyschen Wüste.* Theodor Fischer, Cassell.

Rohlfs, G., 1864–1865. Account of a journey across the Atlas Mountains and through the Oases Tuat and Tidikelt to Tripoli, by way of Ghadames, in the year 1864. *Proceedings of the Royal Geographical Society of London* 9, 312–314.

Ronaldshay, E.O. and Penton, E., 1902. A Journey from Quetta to Meshed via the Nushkisistan Trade-Route. *Geographical Journal* 20, 70–85.

Rothrock, G.A., 1967. Jean-Baptiste Tavernier: Merchant Extraordinary. *History Today* 17, 743–749.

Saalfeld, W.K. and Edwards, G.P., 2010. Distribution and abundance of the feral camel (*Camelus dromedarius*) in Australia. *Rangeland Journal* 32, 1–9.

Sabini, J., 1981. *Armies in the Sand. The Struggle for Mecca and Medina.* Thames and Hudson, London.

Sackville-West, V., 1926. *Passenger to Teheran.* Hogarth Press, London.

Sala, R. 2017., The domestication of camel in the literary, archaeological and petroglyph records. *Journal of Arid Land Studies* 26, 205–211.

Satia, P., 2008. *Spies in Arabia: the Great War and the Cultural Foundations of Britain's Covert Empire in the Middle East.* Oxford University Press, Oxford.

Sattin, A., 2003. *The Gates of Africa. Death, Discovery and the Search for Timbuktu.* HarperCollins, London.

Sazmand, A., Joachim, A. and Otranto, D., 2019. Zoonotic parasites of dromedary camels: so important, so ignored. *Parasites & Vectors* 12, 1–10.

Schmidt-Nielsen, K., 1964. *Desert Animals. Physiological Problems of Heat and Water.* Clarendon Press, Oxford.

Schulz, U., Tupac-Yupanqui, I., Martínez, A., Méndez, S., Delgado, J.V., Gómez, M., Dunner, S. and Cañón, J., 2010. The Canarian camel: a traditional dromedary population. *Diversity* 2, 561–571.

Searle, M., 2019. Exploration of the Oman–UAE Mountains. In *Geology of the Oman Mountains, Eastern Arabia*. Springer, Cham, 27–47.

Seabrook, W., 1928. *Adventures in Arabia: Among the Bedouins, Druses, Whirling Dervishes and Yezideeh Devil-worshippers*. George G. Harrap & Company, London.

Seaver, G., 1952. *Francis Younghusband, Explorer and Mystic.* John Murray, London.

Seifu, E., 2019. Camels in Botswana: Herd dynamics and future development implications. *Botswana Journal of Agriculture and Applied Sciences* 13, 91–106.

Seland, E.H., 2015. Camels, camel nomadism and the practicalities of Palmyrene caravan trade. *ARAM Periodical* 27(1–2), 45–54.

Seton-Karr, H.W., 1898. Further discoveries of ancient stone implement in Somaliland. *Journal of the Anthropological Institute of Great Britain and Ireland* 27, 93–95.

Shaw, W.B.K., 1929. Darb el Arbain. *Sudan Notes and Records* 12, 63–71.

Shaw, W.B.K., 1945. *Long Range Desert Group: The Story of Its Work in Libya, 1940–1943.* Collins, London.

Sim, K., 1981. *Jean Louis Burckhardt. A Biography.* Quartet Books, London.

Skrine, C., 1953. Obituary: Sven Hedin. *Geographical Journal* 119, 252–253.

Sluglett, P., 2004a. Leachman, Gerard Evelyn (1880–1920). *Oxford Dictionary of National Biography Online.*

Sluglett, P., 2004b. Shakespear, William Henry Irvine (1878–1915). *Oxford Dictionary of National Biography Online.*

Smith, A.D., 1896. Expedition through Somaliland to Lake Rudolf. *Geographical Journal* 8, 120–137.

Smith, R.M. and Porcher, E.A., 1864. *History of the Recent Discoveries at Cyrene: Made During an Expedition to the Cyrenaica in 1860–61, Under the Auspices of Her Majesty's Government.* Day, London.

Snouck Hurgronje, C., 1889. *Bilder aus Mekka.* Brill, Leiden.

Spittler G., 1996. Explorers in transit: Travels to Timbuktu and Agades in the nineteenth century, *History and Anthropology* 9, 231–253.

Stark, F., 1942. *Letters from Syria*, John Murray, London.

Steel, J.H., 1890. *A Manual of the Diseases of the Camel and of His Management and Uses.* Lawrence Asylum Press, Madras.

Stephens, J.L., 1839. *Incidents of Travel in Egypt, Arabia Petraea, and the Holy Land.* W. and R. Chambers, Edinburgh.

Sykes, P., 1933. *A History of Exploration.* Routledge, London.

Tabachnick, S.E., 2004. Doughty, Charles Montagu (1843–1926). *Oxford Dictionary of National Biography Online.* https://www.oxforddnb.com/view/10.1093/ref:odnb/9780198614128.001.0001/odnb-9780198614128-e-32866. (accessed August 2021).

Taquet, P., 2007. On camelback: René Chudeau (1864–1921), Conrad Kilian (1898–1950), Albert Félix de Lapparent (1905–1975) and Théodore Monod (1902–2000), four French geological travellers cross the Sahara. *Geological Society, London, Special Publications* 287, 1183–1190.

Tavernier, J.B., 1677. *The Six Travels of John Baptista Tavernier, Baron of Aubonne, Through Turkey and Persia to the Indies, During the Space of Forty Years.* R.L. and M. Pitt, London.

Teka, T., 1991. Introduction: The dromedary in the East African countries: Its virtues, present conditions and potentials for food production. *Nomadic Peoples* 29, 3–9.

Thesiger, W., 1934. Review. *Geographical Journal* 84, 527–528.

Thesiger, W., 1935. The Awash River and the Aussa Sultanate. *Geographical Journal* 85, 1–19.

Thesiger, W., 1948. Across the Empty Quarter. *Geographical Journal* 111, 1–19.

Thesiger, W., 1949. A further journey across the Empty Quarter. *Geographical Journal* 113, 21–44.

Thesiger, W., 1950. Desert borderlands of Oman. *Geographical Journal* 116, 137–168.

Thesiger, W., 1959 and 1960. *Arabian Sands.* Longmans, London.

Thesiger, W., 1960. Obituary of H. St-J. Philby. *Geographical Journal* 126, 563–566.

Thesiger, W., 1987. *The Life of my Choice.* Collins, London.

Thomas, B., 1931. *Alarms and Excursions in Arabia.* George Allen and Unwin Ltd, London.

Thomas, B., 1932. *Arabia Felix: Across the Empty Quarter of Arabia.* Jonathan Cape, London.

Thomas, L., 1924. *With Lawrence in Arabia.* Hutchinson, London.

Thomas, L., 1936. The explorer of the Rub' Al Khali. In L. Thomas, *The Untold Story of Exploration*, Harrap, London, 145–158.

Thomas, N.J. and Hill, J., 2009. Explorations in the Libyan Desert: William J. Harding King. In Ryan J.R. and Naylor, S. (eds.) *New Spaces of Exploration: Geographies of Discovery in the Twentieth Century*, I.B. Tauris, London, 78–104.

Thompson, J., 2004. Palgrave, William Gifford (1826–1888). *Oxford Dictionary of National Biography Online.*

Thorn, D.M., 2007. *The Four Seasons of Cyrene: The Excavation and Explorations in 1861 of Lieutenants R. Murdoch Smith, RE and Edwin A. Porcher, RN.* L'Erma di Bretschneider, Rome.

Tidrick, K., 1990. *Heart Beguiling Araby. The English Romance with Arabia.* I.B. Tauris, London.

Tilho, J., 1910. The French Mission to Lake Chad. *Geographical Journal* 36, 271–286.

Tilho, J., 1910–11. *Documents Scientifiques de la Mission Tilho, 1906–1909.* Imprimerie Nationale, Paris.

Tilho, J., 1920. The exploration of Tibesti, Erdi, Borkou, and Ennedi in 1912–1917: A mission entrusted to the author by the French Institute. *Geographical Journal* 56, 81–99, 161–183, 241–263.

Tomlinson, G.J.F., 1945. Sir Hanns Vischer, C.M.G., C.B.E. *Nature* 155, 446–447.

Townsend, J., 1993. Some reflections on the life and career of Sir Percy Cox, GCMG, GCIE, KCSI. *Asian Affairs* 24, 259–272.

Tuson, P., 2014. *Western Women Travelling East 1716–1916.* The Arcadian Library in association with the Oxford University Press, Oxford.

Tylden, G., 1959. The Camel Corps and the Nile Campaign of 1884–85. *Journal of the Society for Army Historical Research* 37(149), 27–32.

Uerpmann H.-P., 2012. Archaeozoology of camels in south-eastern Arabia. In E. M. Knoll and P. A. Burger (eds), *Camels in Asia and North Africa. Interdisciplinary Perspectives on their Significance in Past and Present.* Austrian Academy of Science Press, Vienna.

Ure, J., 2010. Harry St John Philby. In R. Hanbury-Tenison (ed). *The Great Explorers*. Thames and Hudson, London, 215–219.

Vischer, H., 1909. A journey from Tripoli across the Sahara to Lake Chad. *Geographical Journal* 33, 241–264.

Vischer H., 1910. *Across the Sahara from Tripoli to Bornu*. Arnold, London.

Visher, S.S., 1948. Memoir to Ellsworth Huntington, 1876–1947. *Annals of the Association of American Geographers* 38, 38–50.

Von Dumreicher, A., 1931. *Trackers & Smugglers in the Deserts of Egypt*. Methuen, London.

Von Höhnel, L., 1938. The Lake Rudolf Region Its Discovery and Subsequent Exploration, 1888–1909. Part I. *Journal of the Royal African Society* 37(146), 21–45.

Von Heuglin, T., 1865. *Die Tinne'sche Expedition im westlichen Nil-Quellgebiet: 1863 und 1864; nebst ethnographischen, zoologischen und kartographischen Anhängen und einer Originalkarte*. Perthes, Gotha.

Wallach, J., 1996. *Desert Queen*. Weidenfeld and Nicholson, London.

Wallin, G.A., 1854. Narrative of a journey from Cairo to Medina and Mecca, by Suez, Araba', Tawila, al-Jauf, Jubbe, Hail and Nejd, in 1845. *Journal of the Royal Geographical Society of London* 24, 15–207.

Walpole, R., 1820. *Travels in Various Countries of the East; being a Continuation of Memoirs relating to European and Asiatic Turkey, &c*. Longman, Hurst, Rees, Orme, and Brown, London.

Ward, P., 1987. *Travels in Oman*. Oleander Press, Cambridge.

Wardeh, M.F., 2004. Classification of the dromedary camels. *Journal of Camel Science* 1, 1–7.

Watson, E.E., Kochore, H.H. and Dabasso, B.H., 2016. Camels and climate resilience: adaptation in northern Kenya. *Human Ecology* 44, 701–713.

Weigall, A.E.P.B., 1909. *Travels in the Upper Egyptian Deserts*. W. Blackwood and Sons, Edinburgh and London.

Weigall, A.E.P.B. 1911a. *The Life and Times of Akhnaton: Pharaoh of Egypt*. W. Blackwood and Sons, Edinburgh and London.

Weigall, A.E.P.B., 1911b. *The Treasury of Ancient Egypt: Miscellaneous Chapters of Ancient Egyptian History and Archaeology*. W. Blackwood and Sons, Edinburgh and London.

Weigall, A.E.P.B., 1914. *The Life and Times of Cleopatra, Queen of Egypt*. Putnam, London.

Welland, M., 2015. *The Desert. Lands of Lost Borders*. Reaktion Books, London.

Wellard, J., 1964. *The Great Sahara*. Hutchinson, London.

Wellby, M.S., 1900. King Menelek's dominions and the country between Lake Gallop (Rudolf) and the Nile Valley. *Geographical Journal* 16, 292–304.

Wells, W., 1871. *The Heroine of the White Nile: Or, What a Woman Did and Dared. A Sketch of the Remarkable Travels and Experiences of Miss Alexandrine Tinné*. Carlton & Lanahan, New York.

Wellsted, J.R., 1836. Observations on the coast of Arabia between Ras Mohammed and Jiddah. *Journal of the Royal Geographical Society* 6, 51–96.

Wellsted, J.R., 1838. *Travels in Arabia*. 2 volumes. John Murray, London.

Williams, J.G., 1981. Forth, Nowell Barnard (1879–1933). *Australian Dictionary of Biography*, Melbourne University Press.

Willink R.J., 2007. *The Fateful Journey. The expedition of Alexine Tinne and Theodor von Heuglin in Sudan (1863–1864). A study of their travel accounts and ethnographic collections.* Amsterdam University Press, Amsterdam.

Wilson, R.T., 1984. *The Camel.* Longman, London and New York.

Wilson, R.T., 2008. Perceptions and problems of disease in the one-humped camel in southern Africa in the late 19th and early 20th centuries. *Journal of the South African Veterinary Association* 79, 58–61.

Wilson, R.T., 2012. The one-humped camel in Southern Africa: Imports to and use in South West Africa/Namibia. *Journal of Camel Practice and Research* 19, 1–6.

Wilson, R.T., 2013. The one-humped camel in Southern Africa: use in police, postal service and tourism in Botswana, c.1900–2011. *Botswana Notes and Records* 45, 180–188.

Wilson, R.T., 2016. The one-humped camel in the Anglo-Egyptian military campaigns in Sudan, 1885–1926. *Journal of Camel Practice and Research* 23, 193–205.

Winkler, S., 2004. Vischer, Sir Hanns (1876–1945). *Oxford Dictionary of National Biography.* https://doi.org/10.1093/ref:odnb/40720. (accessed August 2021).

Winstone, H.V.F., 1976. *Captain Shakespear.* Jonathan Cape, London.

Winstone, H.V.F., 1982. *Leachman. O.C. Desert.* Quartet Books, London.

Winstone, H.V.F., 2003. *Lady Anne Blunt: A Biography.* Barzan Publishing, Manchester.

Woodward, A., 1961. *Camels and Surveyors in Death Valley.* Desert Printers, Inc., Palm Desert, California.

Wright, J., 2006. 'An unworn carpet of the Earth: European travellers in the central Sahara, eighteenth to twentieth centuries. In D. Mattingley, S. McLaren, E. Savage, Y. al-Faswati and Khaled Gadgood (es). *The Libyan Desert. Natural Resources and Cultural Heritage.* Society for Libyan Studies, London, 305–313.

Wright, J., 2007. *The Trans-Saharan Slave Trade.* Routledge, London.

Wyatt, E., 2016. *In the Paths of Dangerous Fame.* Independently published.

Wynn, A., 2004. *Persia in the Great Game: Sir Percy Sykes, Explorer, Consul, Soldier, Spy.* John Murray, London.

Xue, Y., Li, J. and Li, D., 2021. The wild camel (*Camelus ferus*) in China: current status and conservation implications. *Journal for Nature Conservation*, p.125979.

Yilmaz, O., Ertugrul, M. and Wilson, R.T., 2011. The domestic livestock resources of Turkey: camels. *Journal of Camel Practice and Research* 18, 21–24.

Zhu, S., Zimmerman, D. and Deem, S.L., 2019. A review of zoonotic pathogens of dromedary camels. *EcoHealth* 16, 356–377.

Zimmermann, M., 1900. La France dans le Sahara. L'occupation d'In Salah. La mission Foureau-Lamy. *Annales de Géographie* 9, 180–182.

Zwemer, S.M., 1902. Three journeys in northern Oman. *Geographical Journal* 19, 54–64.

# INDEX